Building for Tomorrow: Global Enterprise and the U.S. Construction Industry

Committee on the International Construction Industry
Building Research Board
Commission on Engineering and Technical Systems
National Research Council

NATIONAL ACADEMY PRESS
Washington, D.C. 1988

National Academy Press 2101 Constitution Avenue, NW Washington, DC 20418

NOTICE: The project that is the subject of this report was approved by the Governing Board of the National Research Council, whose members are drawn from the councils of the National Academy of Sciences, the National Academy of Engineering, and the Institute of Medicine. The members of the committee responsible for the report were chosen for their special competences and with regard for appropriate balance.

This report has been reviewed by a group other than the authors according to procedures approved by a Report Review Committee consisting of members of the National Academy of Sciences, the National Academy of Engineering, and the Institute of Medicine.

The National Academy of Sciences is a private, nonprofit, self- perpetuating society of distinguished scholars engaged in scientific and engineering research, dedicated to the furtherance of science and technology and to their use for the general welfare. Upon the authority of the charter granted to it by the Congress in 1863, the Academy has a mandate that requires it to advise the federal government on scientific and technical matters. Dr. Frank Press is president of the National Academy of Sciences.

The National Academy of Engineering was established in 1964, under the charter of the National Academy of Sciences, as a parallel organization of outstanding engineers. It is autonomous in its administration and in the selection of its members, sharing with the National Academy of Sciences the responsibility for advising the federal government. The National Academy of Engineering also sponsors engineering programs aimed at meeting national needs, encourages education and research, and recognizes the superior achievements of engineers. Dr. Robert M. White is president of the National Academy of Engineering.

The Institute of Medicine was established in 1970 by the National Academy of Sciences to secure the services of eminent members of appropriate professions in the examination of policy matters pertaining to the health of the public. The Institute acts under the responsibility given to the National Academy of Sciences by its congressional charter to be an adviser to the federal government and upon its own initiative, to identify issues of medical care, research, and education. Dr. Samuel O. Thier is president of the Institute of Medicine.

The National Research Council was established by the National Academy of Sciences in 1916 to associate the broad community of science and technology with the Academy's purposes of furthering knowledge and of advising the federal government. Functioning in accordance with general policies determined by the Academy, the Council has become the principal operating agency of both the National Academy of Sciences and the National Academy of Engineering in providing services to the government, the public, and the scientific and engineering communities. The Council is administered jointly by both Academies and the Institute of Medicine. Dr. Frank Press and Dr. Robert M. White are chairman and vice-chairman, respectively, of the National Research Council.

This report was supported by the Technology Agenda Program of the National Academy of Engineering and funded under the following agreements between the indicated federal agency and the National Academy of Sciences: U.S. Trade and Development Program, International Development Cooperation Agency Grant Agreement TDP 7712561; National Science Foundation Grants No. MSM-8612738 and MSM-8612783 under Master Agreement No. 8618641; Department of the Interior Bureau of Reclamation Grant Agreements No. 6-FG-81-10310 and 7-FG-81-11950, U.S. Department of Agriculture Forest Service Grant No. 87-G-050.

LIBRARY OF CONGRESS CATALOG CARD NUMBER 88-61728
INTERNATIONAL STANDARD BOOK NUMBER 0-309-03937-1

Cover illustration by Tom Adams.

Printed in the United States of America

BUILDING RESEARCH BOARD
(1987–1988)

COMMITTEE ON THE INTERNATIONAL CONSTRUCTION INDUSTRY

JOHN W. FISHER, Fritz Engineering Laboratory, Lehigh University, Bethlehem, Pennsylvania, *Chairman*
DAVID P. BILLINGTON, School of Engineering and Applied Science, Princeton University, New Jersey
ARTHUR J. FOX, *Engineering News Record*, New York, New York
DONALD G. ISELIN, Santa Barbara, California
ARNOLD K. JONES, Cary, North Carolina
MICHAEL MACCOBY, Project on Technology, Work, and Character, Washington, D.C.
HENRY L. MICHEL, Parsons Brinckerhoff, Inc., New York, New York
FRED MOAVENZADEH, Center for Construction Research and Education, Department of Civil Engineering, Massachusetts Institute of Technology, Cambridge
WILLIAM W. MOORE, Dames and Moore, San Francisco, California
LOUIS J. MULKERN, RMD Associates, Washington, D.C.
JOHN C. RICHARDS, Government and International Affairs, M. W. Kellogg Company, Hilton Head, South Carolina
JOHN H. WINKLER, Skidmore, Owings and Merrill, New York, New York

Liaison Representatives

STANLEY BEAN, Forest Products and Harvesting Research, Forest Service, U.S. Department of Agriculture, Washington, D.C.
MARY SAUNDERS, Capital Goods and International Construction, U.S. Department of Commerce, Washington, D.C.
FRANK A. DIMATTEO, U.S. Army Corps of Engineers, Washington, D.C.
CHARLES M. HESS, U.S. Army Corps of Engineers, Washington, D.C.
CHRISTIAN R. HOLMES, International Development Corporation Agency, Washington, D.C.
BETSY HORSMON, Tennessee Valley Authority, Washington, D.C.

THEODORE LETTES, Small Business Technology, U.S. Department of Commerce, Washington, D.C.
RICHARD B. SELF, Executive Office of the President, Washington, D.C.
DARRELL WEBBER, Bureau of Reclamation, U.S. Department of the Interior, Denver, Colorado

Advisers to the Committee

FRANK BOSWORTH, Virginia Polytechnic Institute and State University, Blacksburg
MARION C. DIETRICH, Corporation for Innovation Development, Indianapolis, Indiana
JOHN W. FONDAHL, Stanford University, California
EDGAR J. GARBARINI, Bechtel Group, Inc., San Francisco, California
THOMAS P. GUERIN, JR., Construction/Project Finance, BAII Banking Corporation, New York, New York
H. PETER GUTTMANN, HPG Associates, Inc., Washington, D.C.
GEORGE S. JENKINS, Consultation Networks, Inc., Washington, D.C.
JOHN T. JOYCE, International Union of Bricklayers and Allied Craftsmen, Washington, D.C.
FREDERICK KRIMGOLD, Virginia Polytechnic Institute and State University, Alexandria
NICHOLAS LUDLOW, Development Bank Associates, Inc., Washington, D.C.
RAY MARSHALL, LBJ School of Public Affairs, University of Texas, Austin
ALFRED T. MCNEILL, Turner Construction Company, New York, New York
RICHARD TUCKER, Dames and Moore, Bethesda, Maryland
RICHARD L. TUCKER, Construction Industry Institute, University of Texas, Austin
FRANK M. WARREN, JR., Construction Management Consultant, Charlotte, North Carolina
JOHN WISNIEWSKI, Export-Import Bank, Washington, D.C.
RICHARD N. WRIGHT, Center for Building Technology, National Bureau of Standards, Gaithersburg, Maryland

Observers

JESSE AUSUBEL, National Academy of Engineering, Washington, D.C.
MARLENE R. B. BEAUDIN, National Academy of Engineering, Washington, D.C.
WILLIAM BEDDOW, Caterpillar, Inc., Washington, D.C.
LYNN S. BEEDLE, Lehigh University, Bethlehem, Pennsylvania
TERRY CHAMBERLIN, Associated General Contractors, Washington, D.C.
MARK CHALPIN, National Constructors Association, Washington, D.C.
ROBERT GOLD, Arlington, Virginia
WILLIAM PETERSON, Construction Industries Manufacturers Association Washington, D.C.
CHARLES PINYAN, *International Construction Week,* New York, New York
MARTIN J. THIBAULT, Bureau of Reclamation, U.S. Department of the Interior, Washington, D.C.

Foreword

The U.S. construction industry plays a crucial role in the United States by supplying the structures that house and facilitate virtually all other economic and social activity. This industry has a historic role abroad as well, not only through its direct exports of U.S. goods and services, but also through its leadership in opening opportunities for other U.S. business and for intellectual exchange that improves international understanding. Reports of declining work by U.S. firms abroad and increasing penetration of foreign firms into the domestic construction market are therefore troubling.

Although only a small fraction of the U.S. construction industry is actively involved in the international market, this participation yields a broad range of intangible benefits that go beyond any direct effect on the U.S. trade balance or other economic statistic. These benefits include better knowledge of foreign firms' capabilities and business practices, enhanced skills development through exposure to foreign cultures and management styles, and increased understanding of technical problems arising from construction undertaken in diverse physical and social conditions.

The reasons given for deterioration of the U.S. construction industry's competitive position in an increasingly global marketplace are varied and complex, but the importance of technological leadership is widely recognized. These issues alone would justify an appraisal of the competitiveness of the U.S. construction industry.

However, in requesting the Building Research Board to undertake this study, the National Academy of Engineering had more in mind: Emerging technologies in several fields offer the promise of significant advances in infrastructure and building, at a time when there is growing recognition of the need to renew and enhance these facilities here and abroad. The opportunities presented to U.S. industry by this convergence of capability and need are substantial. The Academy requested this study as one element of a broader effort to identify these opportunities and contribute to the public debate about such issues.

The Academy wishes to thank the National Science Foundation, U.S. Army Corps of Engineers, Bureau of Reclamation, International Cooperation Agency, and Forest Service for joining in the financial support of this study.

Robert M. White
President
National Academy of Engineering

Acknowledgments

This study was conducted in two stages by committees under the chairmanships of William Moore, of Dames and Moore, and John Fisher, of Lehigh University. These committees and their chairmen deserve the particular appreciation of the Building Research Board (BRB) and the National Academy of Engineering (NAE) for their substantial work on this study. Financial support by the U.S. Army Corps of Engineers, Bureau of Reclamation, International Cooperation Agency through its Trade and Development Program, and Forest Service, combined with NAE's initiating funds, demonstrate the government's broad concern about the U.S. construction industry's international competitiveness and the importance of the committee's work.

The committees were ably supported by John P. Eberhard, former Director of the BRB, under whose guidance the study was conducted and who took a major part in preparation of this report. Andrew C. Lemer, currently Director of the BRB, also participated substantially in the report's preparation. Special thanks are due to Joann Curry for her outstanding work on the final manuscript.

Contents

Executive Summary

Construction of housing, other buildings, civil works, and utilities (highways, sewer and water supply systems, railroads, telephone, gas and electric systems) accounts for about 10 percent of the world's total output of goods and services, and well over half of total domestic investment. Buildings and other constructed facilities influence the efficiency of a wide range of economic and social activities, and the productivity of nations.

THE CONSTRUCTION MARKET IN THE UNITED STATES AND ABROAD

Construction is important to the United States. Leaving aside the related industries that produce and transport the materials and equipment of construction, new building constitutes roughly 9 percent of the Gross Domestic Product (GDP)* of the United States and employs 5.5 million people, making the industry the largest single

*Gross National Product (GNP) is a measure of the total value of a nation's output, and includes personal and government expenditures on goods and services and investment, both domestic and foreign. Gross Domestic Product (GDP), which does not include foreign consumption and investment by domestic enterprises, is used as an indication of economic activity within the nation.

component in national accounts. The United States, with an annual domestic construction volume of $330 billion to $390 billion, is about 25 percent of the world total (see Chapter 1).

Foreign companies working in the United States in 1986 accounted for 1 to 2 percent of that amount. Industry observers are concerned that this as yet small penetration of foreign firms may signal the decline of another industry we cannot afford to lose.

Much of the world's construction involves small facilities built by small firms, but a significant portion is undertaken by large firms in international competition. Available data suggest the total volume of international bid construction awarded in 1986 exceeded $74 billion for 250 major firms from more than a dozen countries. U.S. contractors captured $22.6 billion of this work, about 31 percent of the total (see Chapter 2). However, construction by U.S. firms abroad has declined by more than 40 percent since 1983, due both to smaller total construction volumes and declining market share.

In addition to actual construction, design and construction management services represent an increasingly important business in postindustrial economies. U.S. design firms (architects, engineers, and related professions whose markets derive from construction) captured about 26 percent of the estimated $3.5 billion international market in 1986. Again, this is a decline from the 1982 peak of 36 percent.

International construction has domestic importance beyond the contributions to national income that the figures reflect. Many of the 800 U.S. producers of construction equipment export machinery to some 150 countries, often following the lead of U.S. constructors who opened the way. Other types of companies may follow as well, taking advantage of designers' and builders' propensity to specify and use the equipment and materials they know best. However, annual U.S. exports of construction equipment have declined by two-thirds since 1978 to about $2 billion. Employment in the industry has declined similarly.

SOURCES OF CHALLENGES

Declining foreign market shares excite concern about the international competitiveness of the U.S. construction industry. While the reasons for decline lie partly in decreased construction, particularly by those countries like Saudi Arabia where U.S. firms have enjoyed a special relationship, industry leaders cite other problems that hamper U.S. firms abroad and reduce their competitive edge at home:

- Some countries restrict foreign competition for domestic projects.
- Until recently, currency exchange rates made using U.S. firms relatively more expensive.
- Costs to support U.S. professionals in foreign assignments are kept high by U.S. individual and corporate income tax policies.
- U.S. antiboycott and business practice laws restrict U.S. firms' abilities to operate within the foreign business climate of some countries.
- The technological advantages of U.S. firms have been slowly eroded by increasingly capable foreigners. Some of these foreign competitors are based in countries where lower wage scales give additional advantage.

In contrast to other industrialized nations, the United States has no coordinated policy or single government agency to foster international sales of U.S. design and construction expertise. Companies based in industrialized European countries as well as newly industrializing countries in Asia and Latin America have increased their market shares in international construction, often with the aggressive support of their governments (see Chapter 3).

The strategic importance of construction-related export opportunities is reflected in the U.S. Trade and Development Program's support of U.S. firms conducting project feasibility studies. Knowledge gained in the feasibility study can enhance the firm's chances of successfully competing for the much larger construction project. However, this support is much less than many other governments have chosen to provide their nationals, and U.S. suppliers frequently find themselves at a distinct competitive disadvantage.

PRIVATE PRACTICES

International design and construction are dominated by a relatively small and select group of firms. Reliable data are limited, but the Committee on the International Construction Industry estimates that the top 30 construction firms worldwide perform 50 to 60 percent of work available for international competition, and virtually the entire market is captured by 250 major international firms. Of the top 400 U.S. contractors and top 500 U.S. design firms listed in *Engineering News Record* magazine in 1987, 54 construction firms and approximately 200 architecture/engineering firms are actively seeking or conducting international work.

In recent years large amounts of foreign investment capital have entered the United States. For example, the *Los Angeles Times* indicates that increasingly large numbers of downtown office buildings in Los Angeles are foreign owned or controlled. Foreign ownership often opens opportunities for foreign firms to participate in design and construction.

In addition to the direct capture by foreign firms of 1 to 2 percent of the U.S. domestic market, foreign firms are purchasing ownership shares in U.S. construction and design firms or are forming strong associations that may obscure the true volume of foreign participation. From 1978 through 1983, the numbers of foreign design and construction firms forming U.S. affiliations (including purchase of ownership) grew at annual rates of 7.7 percent and 12.8 percent, respectively. Total revenues of these companies, while still less than 2 percent of the total U.S. construction market, grew at an annual rate of 35 percent. Japanese construction volume in the United States reached more than $1.5 billion in 1985.

RESPONSES TO CHALLENGE

The challenges posed by the declining U.S. share of foreign construction and increasing foreign penetration of U.S. domestic markets are substantial. Appropriate response must be balanced among the companies operating in the global marketplace, educational and professional institutions that prepare and support U.S. professionals within these companies, and government policy and institutional support that can motivate and strengthen the private sector. A partnership of diverse interests in the U.S. construction industry is needed to focus resources in research and development (R&D), professional training, and government programs.

RESEARCH AND DEVELOPMENT AND INNOVATION

Research and development and subsequent innovation have been shown in many fields to be valuable elements of competitive advantage. While statistics for construction in general and the U.S. construction industry in particular are limited, they suggest that the U.S. construction industry has fallen behind its competition in its efforts to maintain technology leadership (see Chapter 4): Other countries are putting more effort into construction research and development, backing longer-range research efforts, and actively encouraging the adoption of useful research results to bring innovation to the construction industry.

Japan may be the leader in these efforts. That nation's Ministry of Construction sets national policies on behalf of the construction industry. One of its major policy decisions was to encourage private firms to establish R&D capability comparable to that found in the United States, primarily at universities. As a result of this government policy, more than 20 of the largest firms in Japan now invest 1 percent of their sales in R&D. All have well-equipped campus-like research centers. Research is integrated throughout their operating divisions and has become a major marketing tool for them. Their research programs include a wide spectrum of short-term and long-range projects over a range of technical subjects.

In contrast, total R&D spending on design and construction in the United States has been estimated to be about $1.2 billion annually, only 0.39 percent of the sector's $312 billion of sales in 1984. Compared with R&D spending by other mature industries in the United States (e.g., appliances at 1.4 percent, automobiles at 1.7 percent, or textiles at 0.8 percent), construction industry support of R&D is sorely lagging. Contractors, architects, and engineers as a group invest less than 0.05 percent of sales in R&D, a fraction of the amount they spend on liability insurance alone.

The complex reasons for this lagging effort include the inability of the many firms that make up the U.S. industry to mobilize sufficient resources individually or to consolidate their efforts effectively to support meaningful research or to capture the commercial benefits that may result. The research that is done is concentrated within the universities and is often slow to have an impact on practices in a very competitive industry that is necessarily wary of commercial risk and legal liability.

Yet, rapid advances in technologies now emerging from research laboratories around the world suggest that after decades of relative technological stability, an era of technological ferment, unprecedented in the construction industry, is fast approaching. Leading these developments is the introduction of computer hardware and software into all facets of design and construction. Innovations in construction are likely to result from new inventions emerging from R&D laboratories working in photonics, biotechnology, new materials, microelectronics, and other fields (see Chapter 6).

EDUCATION AND TRAINING FOR GLOBAL ENTERPRISE

Experience working abroad and working with foreign firms at home suggests that key elements for professional success in international design and construction include a strong technical base, understanding of design, understanding of the intimate connection between technology and culture, and experience in foreign languages and regional studies. These elements are gained through a combination of practical experience and formal education that can never really be considered complete (see Chapter 5).

Civil engineering education in the past two decades has emphasized fundamental studies of mechanics, applied mathematics, and the analysis of structures, with relatively less attention to design as synthesis, to construction as the process of economical building, or to the performance and permanence of civil works as measured in field observations. Years of practice may be required to gain the appreciation of design, construction, and performance needed to compete effectively in the world of business. Changes in programs of formal education could lay a stronger foundation for this appreciation:

- Design should be integrated into the teaching of analysis and related directly to construction and performance.
- Young engineers should be educated in the modern traditions, cultural implications, and international potential of the profession.

Architectural education has been shaped by tradition that gives preeminence to apprenticeship and development of strong intuitive understanding of functional and aesthetic bases for building design, with limited attention to technology. Architectural practice is characterized by the proliferation of small design firms and dependence on specialized consultants to address structural, mechanical, lighting, acoustics, and economic issues. Except for a few large and vertically integrated firms, the profession is ill-prepared for the international market. Again, change in professional training could strengthen our competitive stance:

- Technology and the cultural characteristics of construction in other countries should be added to present programs, including courses that foster understanding of how buildings are actually built, not just the materials and equipment that go into a building.
- Working experience in the use of computers as tools of design and analysis should be enhanced.

- Young architects should be exposed to the growing base of research that can support creative design in unfamiliar situations.

Increasingly, as the U.S. design and construction industries look to greater participation in the global enterprise, engineering and architecture schools, professional societies, and business organizations must look outside themselves to learn how to do business in an international economy. Only through more deliberate exposure to foreign languages, geography, business, and foreign culture will U.S. design professionals gain rapid and effective access to foreign-originated technologies, and develop a strong ability to deal with foreign sources of business opportunity and finance.

GOVERNMENT PROGRAMS AND PRIVATE PARTICIPATION

Many observers cite government procurement practices that discourage innovation as well as tax policies and regulations on foreign business practices as evidence of government's failure to support U.S. industry's competitive position. Despite these very real shortcomings, there are examples of effective government efforts in this area:

- Establishment of the National Science Foundation's National Engineering Research Centers, such as the Center for Advanced Technology for Large Structural Systems (ATLSS) at Lehigh University.
- Growth of the government laboratories, such as the Army's Construction Engineering Research Laboratory (CERL), the Navy's Port Hueneme Civil Engineering Laboratory, the Tyndall Air Force Engineering and Services Research Center, and the National Bureau of Standards' Center for Building Technology.
- U.S. Army Corps of Engineers' grants for major new research efforts at the Massachusetts Institute of Technology and the University of Illinois.

However, more could be done to encourage private response to government efforts and to enhance the linkage between research and practical innovation. The programs of other countries illustrate the value to be gained through true partnership of private and public interests in the U.S. construction industry. This partnership should embrace research and innovation for both domestic productivity and international competitive strength.

For example, projects built with government funds can assume the greater commercial risk involved in adopting innovation, as was demonstrated by the introduction to U.S. transit construction of precast concrete segmental tunnel liners (see Chapter 6). However, government cannot act alone. Mechanisms are needed to encourage private-public cooperation in the U.S. construction industry. Precedents for such cooperation exist (the Three Gorges project described in Case Study 6, for instance), but they have been isolated examples. Professional societies and trade associations do well representing the interests of their members, but there is no ongoing means for bringing the industry's diverse interests together to enhance our competitive stance internationally or to foster research and technological innovation at home. A solid institutional focus is needed, and while a number of existing institutions could play a significant role in creating this focus, a new organization may be required (see Chapter 7).

BUILDING FOR TOMORROW

Within the United States, as in most of the industrial world, there is an opportunity to increase the performance characteristics of those infrastructure systems used to transport people and goods, obtain water, remove wastes, supply energy, and facilitate communications. There is also reason to include those buildings used either for public purposes (e.g., schools and hospitals) or built with public funds (e.g., government offices, court houses, and prisons) as a part of the public works infrastructure. Under this broad definition of infrastructure the United States in 1984 invested 30 percent of its design and construction budgets in these facilities, a total of $102 billion, and other countries are investing as well (see Chapter 6).

Development of advanced infrastructure is a challenge worthy of cooperative international effort. It will be difficult to structure these developments to match the performance requirements of a society utilizing advanced science and technology, and make more than incremental improvements to the present modal technologies. In the developing part of the world, which is experiencing the most rapid urbanization, the challenge is to develop technology appropriate to their requirements rather than to impose solutions produced for industrial nations.

There are two reasons for the United States to do more toward advancing the technology of infrastructure. We would benefit within our own borders from new and higher-performance systems, and we could also enhance the opportunity for marketing our technology on a global basis. This committee recognizes the urgency of maintaining and extending the existing networks of public works that underlie our nation. However, we need also to develop new and higher-performing technologies to enhance our competitive position in the world.

We are faced as a nation with a challenge to build for tomorrow. The strategic and commercial rewards of meeting this challenge will be surpassed only by the rewards of improved quality of life for the citizens of an increasingly global economy.

1 Introduction

The prominent role of construction in the wealth of nations is readily apparent in the buildings and infrastructure facilities that enable much of modern life. Leaving aside the related industries that produce and transport the materials and equipment of construction, new building accounts for roughly 9 percent of the Gross Domestic Product (GDP) of the United States and employs 5.5 million people, making the industry the largest single component in national accounts. Comparisons among nations show that construction tends to account for an increasing share of GDP as per capita incomes rise in early stages of growth, and dominates investment in countries at all levels of development.

Given the scale of construction and its associated design, materials, and equipment businesses within the U.S. economy, there is a surprising lack of detailed statistics and definitive analysis of this sector's structure, performance, and contribution to the nation's growth and development. Knowledge of the construction industry in other countries is poorer still.

The Committee on the International Construction Industry found it necessary to rely on its members' experience and accounts told by others to supplement the meager base of statistical data. The committee found in some cases that these accounts taught informative lessons and made them the basis for the case studies presented herein.

The committee's work reveals a picture that is somewhat impressionistic in nature, based on a combination of these limited statistical data and case studies. The committee found, nevertheless, a need for changes in education, support for research, and enthusiasm for innovation in the construction industry. These changes are needed to enhance both the nation's ability to capture major new opportunities from technological progress that seems likely to alter in basic ways the physical infrastructure of society, and its competitive strength in an increasingly global market for construction services.

THE SCALE OF WORLD CONSTRUCTION

Estimates based on governmental records indicate that the world invests about $1,430 billion every year in the construction of housing, other buildings, civil works, and utilities (highways, water and sewer, railroads, telephones, gas, and electricity), or a little more than 10 percent of the world's GDP (see Table 1). Actual amounts may be even greater. Construction is the largest industry in the world.

As individual countries develop, rising per capita incomes spur growing demand for more and better buildings and infrastructure, and construction accounts for an increasingly significant share of national economic activity. Some evidence suggests that construction's share of economic activity may stabilize or decline at higher levels of development, but the level remains high even in the most advanced country. The U.S. annual domestic construction volume of $330 billion to $390 billion is about 25 percent of the worldwide total.

Much of the world's construction is done by small-scale builders who produce single houses or maintain roads over small areas, using very traditional building materials and methods. Only perhaps one-fifth of the total volume of construction is consistently carried out by large-scale organizations using more modern methods, as well as traditional methods that remain predominant in current practice.

Much of this submarket is in turn limited by political and economic reasons to domestic firms or government agencies using local materials, labor, and design and management services. Total construction undertaken in a fully internationally competitive market in 1986 exceeded $74 billion, or about 5 percent of the world's construction (see Table 2).

This market, while only a fraction of total construction, is nevertheless a big business and it is dominated by a relatively few major

TABLE 1 Comparative Statistics on Building as a Component in National Economies, 1984

Country	GDP[a] ($million)	Building Value Percentage of GDP	Building Value Total ($million)	Source
		Low-Income Economies		
Ethiopia	4,270	3.0	128	
Bangladesh	12,320	5.0	616	UN
Mali	980	3.0	29	
Zaire	4,700	3.0	141	
Burkina Faso	820	3.0	25	
Nepal	2,290	7.0	160	UN
Burma	6,130	3.0	184	
Malawi	1,090	3.0	33	
Niger	1,340	3.0	40	
Tanzania	4,410	3.0	132	UN
Burundi	1,020	4.0	41	UN
Uganda	4,710	3.0	141	
Togo	420	8.0	34	UN
Central African Republic	460	4.0	18	
India	162,280	5.0	8,114	UN
Madagascar	2,380	4.0	95	
Somalia	1,364	4.0	55	
Benin	900	4.0	36	UN
Rawanda	1,600	4.0	64	
China, People's Republic of	281,250	4.0	11,250	
Kenya	5,140	5.0	257	UN
Sierra Leone	900	4.0	36	
Haiti	1,820	4.0	73	
Guinea	2,100	4.0	84	
Ghana	4,485	4.0	179	
Sri Lanka	5,430	7.0	380	UN
Sudan	6,730	4.0	269	
Pakistan	27,730	5.0	1,387	UN
Senegal	2,390	4.0	96	
Afghanistan	3,000	5.0	150	UN
Bhutan	3,000	4.0	120	
Chad	360	4.0	14	
Laos, People's Democratic Republic of	765	4.0	31	
Mozambique	2,000	4.0	80	
Vietnam	18,100	4.0	724	
Total, Low-income economies			25,216	
		Middle-Income Economies		
Mauritania	660	4.0	26	
Liberia	980	7.0	69	UN
Zambia	1,060	4.0	42	UN

Continued

TABLE 1 (Continued)

Country	GDP[a] ($million)	Building Value Percentage of GDP	Building Value Total ($million)	Source
Lesotho	360	4.0	14	
Bolivia	3,610	5.0	181	UN
Indonesia	80,590	6.0	4,835	UN
Yemen Arab Republic	2,940	5.0	147	
Yemen, People's Democratic Republic of	792	4.0	32	
Cote d'Ivoire	6,690	9.0	602	UN
Philippines	32,840	19.1	6,272	CICA
Morocco	13,300	5.0	665	
Honduras	2,840	5.0	142	UN
El Salvador	4,070	5.0	204	UN
Papua New Guinea	2,360	5.0	118	
Egypt, Arab Republic	30,060	5.0	1,503	UN
Nigeria	73,450	5.0	3,673	
Zimbabwe	4,580	3.0	137	UN
Cameroon	7,800	5.0	390	
Nicaragua	2,830	3.0	85	UN
Thailand	41,960	5.0	2,098	UN
Botswana	990	5.0	50	UN
Dominican Republic	4,910	5.0	246	
Peru	18,790	2.0	376	UN
Mauritius	860	5.0	43	
Congo People's Republic	2,010	5.0	101	
Ecuador	9,870	7.0	691	UN
Jamaica	2,380	5.0	119	
Guatemala	9,400	5.0	470	
Turkey	47,460	5.0	2,373	UN
Costa Rica	3,560	5.0	178	
Paraguay	3,870	5.0	194	
Tunisia	6,940	5.0	347	UN
Colombia	34,400	4.0	1,376	UN
Jordan	3,430	12.0	412	UN
Syrian Arab Republic	15,930	5.0	797	UN
Angola	4,000	5.0	200	
Cuba	14,900	6.0	894	
Korea Democratic Republic	23,000	6.0	1,380	
Lebanon	5,300	6.0	318	
Mongolia	1,200	5.0	60	
Total, Middle-income economies			31,857	
Upper-Middle-Income Economies				
Chile	19,760	6.0	1,186	
Brazil	187,130	5.0	9,357	UN
Portugal	19,060	11.4	2,173	CICA

Continued

TABLE 1 (Continued)

Country	GDP[a] ($million)	Building Value: Percentage of GDP	Building Value: Total ($million)	Source
Malaysia	29,280	6.0	1,757	
Panama	4,540	6.0	272	
Uruguay	4,580	5.0	229	UN
Mexico	171,300	15.5	26,552	CICA
Korea, Republic of	83,220	9.4	7,789	CICA
Yugoslavia	38,990	11.0	4,289	UN
Argentina	76,210	6.0	4,573	
South Africa	73,390	6.0	4,403	
Algeria	50,690	6.0	3,041	
Venezuela	47,500	7.0	3,325	UN
Greece	29,550	9.0	2,660	UN
Israel	22,350	10.0	2,235	UN
Hong Kong	30,620	7.0	2,143	
Trinidad and Tobago	8,620	8.0	690	UN
Singapore	18,220	23.9	4,355	CICA
Iran Islamic Republic	157,630	5.0	7,882	
Iraq	27,000	5.0	1,350	
Total, Upper-middle-income economies		90,259		
High-Income Oil Exporters				
Oman	7,680	8.0	614	UN
Libya	30,570	11.0	3,363	UN
Saudi Arabia	109,380	15.0	16,407	UN
Kuwait	21,710	3.0	651	UN
United Arab Emirates	28,840	10.0	2,884	UN
Total, High-income oil exporters			23,919	
Industrial Market Economies				
Spain	160,930	8.0	12,874	UN
Ireland	18,270	9.0	1,644	
Italy	348,380	11.4	39,715	CICA
New Zealand	23,340	4.0	934	UN
United Kingdom	425,370	8.5	36,156	CICA
Belgium	77,630	7.0	5,434	CICA
Austria	64,460	28.0	18,049	CICA
Netherlands	132,600	13.0	17,238	CICA
France	489,380	11.3	55,300	CICA
Japan	1,255,006	23.3	292,416	CICA
Finland	51,230	10.0	5,123	CICA

Continued

TABLE 1 (Continued)

Country	GDP[a] ($million)	Building Value: Percentage of GDP	Building Value: Total ($million)	Source
Germany, Federal Republic of	613,160	14.0	85,842	CICA
Denmark	54,640	10.0	5,464	CICA
Australia	182,170	12.0	21,860	
Sweden	91,880	12.7	11,669	CICA
Canada	334,110	12.0	40,093	
Norway	54,720	12.0	6,566	
United States	3,634,600	9.0	327,114	Census
Switzerland	91,110	14.7	13,393	CICA
Total, Industrial market economies			996,887	
East European Nonmarket				
Hungary	20,150	12.0	2,418	UN
Poland	75,410	11.0	8,295	UN
Albania	2,700	11.0	297	
Bulgaria	56,400	8.0	4,512	UN
Czechoslovakia	127,900	8.0	10,232	
German Democratic Republic	163,700	7.0	11,459	UN
Romania	117,600	10.0	11,760	UN
Soviet Union	1,957,600	11.0	215,336	UN
Total, East European nonmarket			264,309	
World Total	13,027,922		1,432,447	

[a]Gross Domestic Product.

Note: Data are from United Nations' (UN) reports, the Swedish industry report (CICA), and the U.S. Census Bureau.

firms. More than half of the work (in dollar volume) is done by the top 30 contractors. In the United States, which may have a higher proportion of moderately sized firms than other countries, 200 firms (about 1.7 percent of all U.S. constructors) conduct about 85 percent of the business.

THE CHANGING MARKET

The need for construction of new facilities combined with poorly developed domestic construction industries has made developing countries the primary locus of international competition in the past.

TABLE 2 Summary of Estimated Market Structure for International Construction

Market Segment	Annual Amount (U.S. $billion)
Estimated total construction worldwide	1,430[a]
Small scale and restricted	1,140[b]
Modern method and management	290[b]
Restricted or communist bloc projects	216[b]
International construction market	74[c]
Foreign contracts of top 30 constructors	44[c]
Remaining international construction market	30[b]

[a]From United Nations' data and surveys. (See Table 1.)
[b]Committee staff estimates.
[c]International construction week, Engineering News Record, July 20, 1984.

The economic upheaval of oil and commodity price fluctuations and growing debt burdens, however, has slowed construction growth from the average of 6 percent annually between 1967 and 1976 to 1.5 percent in 1983. Construction of large and technically complex projects has come to a virtual standstill in many countries. Some countries are beginning to show signs of slow recovery, but without question the international market remains constricted.

At the same time, larger numbers of firms are competing in this limited market. These firms fall into four categories. First, some firms (typically British, French, Dutch, and Scandinavian) have long experience with construction export and extensive contacts throughout developing countries. This experience has been fostered largely by economic and political opportunity. As a result of former colonial ties, for example, the French construction industry has special access to many countries in West Africa and the Middle East, and the British construction industry to the subcontinent of Asia, Southeast Asia, the Middle East, and East Africa. To some degree the United States has enjoyed such a relationship with Saudi Arabia. The markets these relationships provide are sensitive to changing economic and political environments, but give these firms a distinct competitive advantage.

The second category includes firms based in industrialized countries that have not previously been substantial exporters of construction, but in the past 10 to 15 years have initiated efforts to export their surplus domestic capacity. In countries such as Italy and Japan, reconstruction efforts following World War II created extensive construction industries to meet domestic needs. Such demand is now dropping off sharply, and considerable surplus construction capacity exists in their domestic industries. Companies from these industrialized countries are operating under conditions similar to those of American firms. They have a highly developed technology base, they have sophisticated management and technology products, their financing capability is considerable, and their architectural and engineering fees are comparable to those of the United States. The nature of the competition among these countries is based on the quality of technological ability and the adequacy of financing. Scheduled removal of trade barriers among nations of the European Economic Community (the EEC, or Common Market) will give these firms a domestic market comparable to that of the United States, and competition may be intensified.

Firms based in newly industrialized countries, such as Korea, Brazil, Taiwan, Turkey, India, and the Philippines, constitute a third form of competition. These countries have developed construction capacity as an element of their national economic planning and have invested in export construction capability as a means of raising export income. Firms in these countries are characterized by a developing but limited technical capacity and by relatively low wages. Advanced technology is typically available by license or other arrangement. However, their fundamental basis for competing in international markets is essentially low price, both for construction labor and for professional services. Management skills and technological capability are increasing in these countries at a very rapid rate. There are now few projects upon which the national construction industries of these countries cannot bid competitively. However, where sophisticated technology is required, there continues to be a strong incentive to involve European, Japanese, or U.S. contractors.

Firms based in the developing countries constitute a fourth group of competitors, but they currently do not have the capability to pose a serious commercial threat in world markets. However, in many developing countries emphasis is being placed on the development of a basic local construction industry for import substitution. This emphasis reduces hard currency expenditure for goods and commodities that can be provided domestically, and in many such countries has provided a vehicle for gradual increase in technological capability and labor skills, and investment in industrial capacity. In these countries there is often a strong effort to protect this domestic industry (organizations such as the World Bank have institutionalized a policy preference for utilizing the developing country's own sources of construction). Many developing countries that previously provided opportunities for foreign construction companies are no longer open to the international market.

While the number of U.S. firms that compete in this global market is small compared to the total number of firms in the design and construction business, these firms are generally very large employers (by construction industry standards) and are key players in the international competition. Unlike its foreign competition, the United States has been slow to develop national trade and economic policies in support of international engineering and construction. In this country there is no central policy-coordinating agency for construction, in contrast with much of the rest of the world, where there is a cabinet-level officer who heads a ministry of construction or its equivalent.

Domestically, the construction industry is largely decentralized and generally in a defensive mode. Consolidations are taking place across the industry, with foreign investors buying large interests in some firms, and other firms are closing shop. The design community now finds mergers and acquisitions with firms from other countries commonplace, especially for those firms that became visible by competing in the international arena. While some companies established leadership through control of technology valuable to manufacturing firms that are clients for construction (see Case Study 1), most U.S. international construction companies have grown from an initial specialization in one of the following market segments: electric power generating facilities, highways, mining, refinery facilities, and large dams. Regardless of their origins, however, these firms maintained leadership through technology developments and management skills that are increasingly shared by competitors.

CASE STUDY 1:
TECHNOLOGICAL ADVANTAGE PAYS OFF: M. W. KELLOGG AND THE OIL AND PETROCHEMICAL INDUSTRY

U.S. construction firms have reaped over several decades the benefits of the nation's technological leadership in industry. Some U.S. builders simply followed their long-time domestic manufacturer clients who moved abroad in a quest for larger markets and supply sources. Other firms have played a more active role in developing the technology that opened market opportunities in countries seeking to expand their industrial bases.

Since its formation in the first year of this century, the M. W. Kellogg Company has relied on experienced mechanical and chemical engineering techniques and laboratory research and development to grow as an engineering contractor by creating and improving new processes for the oil and petrochemical industry. A small pipe fabrication and chimney business soon evolved into power piping. The company began experimenting with a new hammer-forge welding technique it learned from German industry, and subsequent development work improved the methods used, providing the basis for entrance into the petroleum field.

Approached by Richard Fleming in 1919 to develop a new oil cracking method, Kellogg hired the inventor and developed his process. Fleming units were installed at several oil refinery units, providing much higher gasoline yields than with conventional equipment.

In the early 1920s oil refineries were converting only 30 percent of their crude oil to gasoline, and the heavy demand for motor gasoline dictated a need for higher recovery. The answer came in 1924 with the introduction by the Cross brothers of a new high-pressure thermal cracking process. Under a special agreement, Kellogg was brought in to help develop the process, and for this purpose a laboratory was set up in 1926. One of the first petroleum laboratories in the country, its studies resulted in the successful commercialization of the Cross process. In the 10 years following, Kellogg built more than 130 Cross units in the United States and abroad. Twenty Cross units were built overseas: five in Argentina; three in England; two each in Japan, Poland, and the Dutch West Indies (Aruba); and one each in Brazil, France, Indonesia, Italy, Mexico, and Portugal.

The Cross process development was followed by further development of thermal processing technology. By 1939 some 45 percent of the crude oil could be converted to motor gasoline.

To study the thermal cracking process in more detail, Kellogg set up a new laboratory in 1931 especially for that purpose. This research work and the related process and mechanical engineering design produced the combination unit concept that made an important contribution to thermal cracking progress. This design was a first step in process integration for improved economy and gave an early impetus to continuous plant process design and larger, more efficient oil refineries.

In cooperation with major oil companies, Kellogg's knowledge of catalytic processing grew toward a major accomplishment—the technical development of fluid catalytic cracking of gas oils. Preliminary studies in this field were a part of an exploratory research program cosponsored with Standard Oil of New Jersey, Standard Oil of Indiana, and the Texas Company. A separate laboratory was established for this work, and by 1938 Kellogg had in operation a continuous fluid moving bed catalytic cracking pilot plant and began an exchange of information that led to the commercialization of the process. The original idea for using a powdered catalyst came from Standard Oil of New Jersey, and Kellogg turned its attention to placing this unique concept in practical application.

In 1941 fluid catalytic cracking was drafted into war service to satisfy the great need for aviation gasoline—before the process had gone beyond the pilot plant stage. Kellogg placed its first fluid unit in operation in 1942 for Standard Oil of Louisiana at Baton Rouge and had 20 units in production when the nation's aviation gasoline program ended in 1944. These early units were built essentially from Kellogg pilot plant data.

Paralleling the pioneering activities in the petroleum area were technical contributions in the field of cryogenics and gas processing. This began in 1937 with a study on oxygen generation to be used in a process producing hydrocarbon liquids from coal. This early research and development work led to the construction in South Africa of the world's first successful large-scale plant producing synthetic oil and gas from coal.

These research and engineering activities provided strong technical positions in gas processing, synthetic fuels, and ethylene production. The extensive basic work carried on through the years has provided a large share of Kellogg's domestic and international business over the past several decades.

Since 1975 when Kellogg was eighteenth on the Engineering News Record (ENR) annual list of 400 contractors (the ENR 400) with a *total contract volume of $1 billion, Kellogg has increased its business volume substantially. The company headed the list of the ENR Top 250 International Contractors for 1984, 1985, and 1986, with total contract volumes in the range of $10.9 billion (1984) and $6.9 billion (1986). The foreign contract volume as a percentage of the total during these three years was 60 to 80 percent.*

This improvement in business volume is believed to be largely attributable to the competitive edge gained through the achievement of strong technical positions in several proprietary processes such as synthetic ammonia, ethylene, and liquified natural and petroleum gas processing. These processes were developed and improved over many years through the continuing activities in Kellogg research and development laboratories and chemical and mechanical engineering groups. Certainly the proprietary position in these processes has contributed in large measure to the high-percentage volume of international business.

Perhaps Kellogg's most impressive technical achievement was the more recent development and commercialization of a radically new process plant that makes possible the production of ammonia in large quantities at significantly reduced cost. The new approach incorporated earlier process and equipment design developments such as higher steam reforming pressure, lower ammonia synthesis pressure, and the use of steam-driven centrifugal compressors instead of reciprocating compressors in all major services. All of these design innovations resulted in greatly improved energy efficiency. Operating costs were reduced appreciably by generating super-heated steam at elevated pressures and using the steam in a series of efficient extraction steps involving both process users and steam turbine drives for all major pumps and compressors.

The first two plants of the new large-scale single-train ammonia plant design were sold to Imperial Chemical Industries (ICI) in England. Within one year, 10 more 600- and 1,000-ton-per-day plants were ordered in Europe, including a third duplicate 1,000-ton-per-day plant for ICI. In the United States, a 600-ton-per-day plant was placed in service in July 1965 for Monsanto in Louisiana, and within one year three more large-scale plants were brought on-stream in Louisiana and Mississippi. These plants cut the cost of producing ammonia in half and sent producers off on a major round of new plant expansion worldwide.

Between 1963 and 1983 the Kellogg worldwide record in large-scale ammonia plants numbered 132, of which 83 are in production outside North and South America. These plants now supply more than half the world's synthetic ammonia, the principal base material for most fertilizers.

A significant technical and commercial breakthrough was also made in the People's Republic of China through the sale of Kellogg's ammonia and urea technology in the early 1970s. Following the signing of the famous Shanghai Communique of 1972, Kellogg set in motion a marketing program in the fall of the year. China's need for nitrogen was well known and Kellogg volunteered to submit a proposal. The initial proposal was rejected because of some open cost features, and a lump sum proposal was later accepted for one ammonia plant. This agreement was followed by a surprising request for two more plants, and a contract for the three plants was signed in June 1973.

Concurrently, the Chinese were negotiating with a Japanese firm for ammonia plants using Kellogg technology. The Japanese received contracts for two plants, providing the Chinese with an excellent benchmark in negotiations with Kellogg, but at the same time giving the company additional revenue for ammonia design know-how.

An even greater surprise followed with a Chinese request for five more ammonia plants, and contracts for these plants were signed in November 1973. At about the same time as the ammonia plant negotiations were under way, Kellogg's Dutch company completed contract negotiations for eight urea plants that use the ammonia and carbon dioxide produced in the ammonia plants at eight different sites in China. The contract value of Kellogg ammonia and urea work in China represented about $500 million in business.

2
U.S. Construction in International Competition

The U.S. construction industry has fared poorly in this difficult climate of stagnant markets and growing competition. A more detailed look at the structure of the U.S. industry and some of its principal competitors in international markets reveals erosion of traditional technological advantages and failures to keep up in developing the skills needed for competition.

Available data indicate that U.S. construction firms in 1986 captured $22.6 billion in new contracts, almost 31 percent of the international export market (see Table 3). This amount represents a decline of more than 40 percent in sales dollars since 1982.

U.S. design firms (engineers, architects, and construction managers whose markets are driven by construction) working internationally often provide some advantage for U.S. construction firms. These firms garnered $917.8 million in 1986 billings, about 26 percent of the market (see Table 4). Again, these figures represent a sharp decline from 1982, when U.S. firms captured 36 percent of a market made fat by the spending of prosperous oil-producing countries.

THE U.S. INDUSTRY

The small number of U.S. firms competing in the global market are generally very large employers (by construction industry standards) and are key players in the international competition.

TABLE 3 International Construction Shares, 1986

Nation of Contractor	Number of Firms	Total Awards $Billion	Total Awards Percentage
American	43	22.6	30.6
Japanese	29	9.4	12.7
Korean	14	2.6	3.5
European	126	33.7	45.5
Italian	35	7.4	10.0
French	18	7.1	9.6
British	17	7.0	9.5
German	17	5.5	7.5
Yugoslavian	6	1.4	1.9
Swiss	5	1.3	1.7
Dutch	7	1.1	1.5
Other	21	2.9	3.9
Turkish	9	2.2	3.0
All other	29	3.4	4.7
Total	250	73.9	100.0

Source: Engineering News Record, July 16, 1987.

Note: Data are based on voluntary responses to a survey.

In 1983, U.S. firms involved in international contracting employed 45,000 Americans and 99,000 people of other nationalities.

Domestically, the construction industry consists of many small firms that respond to externally determined demand. Consolidations are taking place across the industry, with foreign investors buying large interests in some firms, and still other firms are closing shop.

Some people believe that these consolidations and mergers are an attempt by the marketplace to sort the industry into two broad categories: the all-purpose firms (not unlike their Japanese counterparts), and the specialized "boutiques" (small, but highly specialized). Overlying this restructuring of the industry is a constant struggle with a litigious society in which each party to a contract has found itself confronted in a court of law. In such a climate, too many organizations devote energies and management structure largely to minimizing risks, rather than building new markets or applying innovations.

TABLE 4 International Design Shares, 1986

Nationality of Designer	Number of Firms	Foreign Billing $Million	Foreign Billing Percentage
American	49	917.8	25.9
European	106	1,958.4	55.3
British	26	481.4	13.6
French	15	306.3	8.6
German	21	282.1	8.0
Dutch	8	259.3	7.3
Scandinavian	11	227.1	6.4
Swiss	8	174.7	4.9
Other	17	227.9	6.5
Canadian	14	204.0	5.8
Japanese	12	220.5	6.2
Korean	4	54.0	1.5
All other	15	185.1	5.3
TOTAL	200	3,539.9	100.0

Source: Engineering News Record, August 6, 1987.

Note: Data are based on voluntary responses to a survey.

A Short Historic Perspective

Until the Industrial Revolution, construction remained little changed from Roman times. Stone, brick, and timber were used for buildings, and infrastructure was rudimentary.

By the end of the nineteenth century, a "second generation" of essentially urban inventions (structural steel frames, the elevator, electrical systems, sewer and water systems, indoor plumbing, central heating, the telephone, the automobile and highway, and the subway) was ready for worldwide diffusion and installation. Most of the world's construction industry known today came into being to integrate these inventions into individual communities.

After World War I and the subsequent boom and bust periods of the 1920s and 1930s, construction capabilities increased to include the building of national highways, large reclamation projects, and dams for water control and power production. The U.S. Army Corps of Engineers and the Bureau of Reclamation played major roles in shaping and managing such projects. As the country matured so did the industries of construction.

At the end of World War II, the physical restructuring of the world's cities, many of which had been destroyed or damaged by the war, was aided by such major government programs as the Marshall Plan and President Harry Truman's Point Four program for Third World countries. The devastated urban areas of the European continent, the Soviet Union, the Middle East, North Africa, many islands in the Pacific, China, Korea, and Japan were much in need of "construction" and "reconstruction." The United States alone retained relatively undamaged physical facilities, an economic base, and the resources to aid in this global program. During the war, the United States had created the impressive organizational capacity of the military construction arm of the Army Corps of Engineers and the Navy Seabees. With the development of multinational corporations, which became clients for construction projects in other countries, a further incentive was added for other U.S. design and construction firms to move into the international arena.

A "military" component to the Marshall Plan included the placement of U.S. military bases on foreign soil to counter the perceived Soviet threat. Most of the physical infrastructure for these military bases was originally built by the engineering elements of the U.S. armed forces, who were soon superseded by a number of the larger and more aggressively profit-motivated private sector design and construction firms. U.S. engineering and construction firms were employed by European industry to undertake much of the reconstruction work for the private sector as well. In turn, a parallel effort was begun by European firms who were reentering the market following a period of dormancy during the war, and who were adopting many of the techniques and much of the equipment of their U.S. counterparts.

This pattern persisted throughout the 1950s and 1960s, in both military and civilian sectors, first in Korea and then in Vietnam. The phenomenal growth of the South Korean construction industry can be attributed in large part to the close working relationship between the Corps of Engineers and its South Korean counterpart. The Koreans were rapid learners and within a few years had put together a number of large and capable construction companies. These companies became especially prominent and successful in the latter days of the construction boom in Saudi Arabia, and became a very lucrative source of foreign income for Korea. To a much lesser extent, the same pattern was followed in Japan and Taiwan.

The case of Saudi Arabia, and to a lesser extent other oil-producing nations, in the 1970s is a special one and not likely to be repeated. Oil and oil pricing made available an unprecedented amount of capital to Saudi Arabia and its neighbors for imports and construction projects. The Saudis had enjoyed a close relationship with the United States since the early phases of Aramco and during World War II. Because Saudi Arabia did not have any of the requisite technological capability or project management expertise, its national leaders turned to the United States. The result was major participation by U.S. engineering and construction firms—such as Bechtel, Fluor, and Ralph Parsons—in contracts for planning, civil and mechanical engineering design, and some construction management. The U.S. Army Corps of Engineers, familiar with large-scale projects, was designated the overall project manager for many military-related projects, working very closely with the Saudi Arabian authorities. Practically all design and engineering projects were awarded to U.S. design firms, since U.S. specifications were being used. The construction projects were open to international competition. Early involvement in a project usually increases the odds of later work for the design and engineering team (see box), but American design teams cannot ensure that the construction phase will go to American firms when public funding is used. Once the actual construction is under way, the products used in the building can be purchased from a country other than the home country of the design team.

In the past few years the Trade Development Program within the U.S. State Department has provided critical funding for a large number of feasibility studies by U.S. design and construction firms. The financial support is given to those firms whose projects show the prospect of a major return to the economy if they obtain the contract.

Market Structure

Construction, the largest industry in the United States and the major employer, is a relatively disaggregated and volatile market that responds to interest rates and levels of general economic activity. The industry's 1.2 million firms undertake more than $360 billion in contracts each year and employ 5.5 million workers. When the suppliers of materials, machinery, insurance, and design services, and the operation and maintenance of all constructed facilities are added to this total, the overall industry accounts for 17 percent of the U.S. work force. Construction has traditionally made up some 55 to 65 percent of the nation's capital investment.

PROJECT CYCLE FOR MAJOR CONSTRUCTION PROJECTS

A construction project passes through three relatively distinct phases: feasibility analysis, design and engineering, and actual construction. Construction materials and labor account for about 85 percent of typical project costs, with the balance being professional services. As a rule of thumb, feasibility studies are about 1 percent of the total project cost, design and engineering fees are about 10 percent, construction management can run between 2 and 6 percent. Operation and maintenance costs over the 20- to 50-year life of the facility can approach several times the project's initial costs.

While clearly small in scope, feasibility studies can provide an invaluable opening wedge for engineering contracts to follow. Both the World Bank and the Inter-American Development Bank provide for "continuity of work." Under this policy a client can award the design contract to a firm as a follow-up contract to the feasibility phase without reopening the contracting process if the client is satisfied with the earlier work.

Thus, the linkage between feasibility studies and design work can be exploited as a marketing tool. Feasibility studies can be underpriced or financed at generous terms to land the design contract. Consulting engineers may be faced with the challenge of maintaining objectivity when the outcome may influence future opportunities for work.

Once the feasibility study has been accepted by the client, and very often by the financing organization, the design firm is chosen. Given the nature of the project, the design firm may be more heavily oriented toward engineering than architecture, or vice versa. Often the firm has both qualifications. Also at this stage, the project management organization might be chosen depending on the project's complexity and the client's desires.

During the design phase, the detailed working drawings and specifications are prepared for contractors' bids. Depending on the circumstances, a bidders' list based on prequalifications may be prepared. Any firm with the ability to post a construction bond will normally be allowed to bid on public projects. Award of the contract is almost always to the contractor with the lowest price.

Procurement of supplies, materials, building components, mechanical and electrical equipment, and construction labor will be determined by a large number of factors. However, once the construction contract is awarded, it is common practice for the construction firm to procure these items from suppliers within their home country, thus creating an "after market" for replacement parts and additions based on upgraded performance.

TABLE 5 U.S. Construction Market by Sector, 1985 (in $billions of new construction)

Market Sectors	Federal Information[a]	Industry Information[b]
Residential	159	159
Office and commercial	60	60
Institutional	10	10
Hotels and motels	7	7
All other private	8	8
Subtotal	85	85
Industrial	16	54
Electric power	16	20
Other utilities	17	17
Subtotal	49	91
State and local government	50	50
Federal government	12	12
Total	355	397

[a]U.S. Census Bureau data.
[b]Construction Industry Institute (CII) adjustments to data, based on the knowledge of their members. The CII estimates are larger for the industrial market sector and the electric power sector because of "force" accounts, that is work done by the employed staff of industrial firms and therefore not publicly bid or counted in census data which are largely based on records of building permits.

The design and construction industry is organized around market sectors that are widely different in terms of the type of customer, the method of financing, the work force used, and even the level of technology. Table 5 presents a common way of indicating these market sectors.

The "residential" (housing) design and construction sector is primarily made up of smaller independent builders. The largest home builders and developers in the United States have no more than $2 billion of this $159 billion market. The balance of the business is conducted by the thousands of firms with fewer than 100 employees.

The manufactured housing industry has grown to capture a larger share of this market since World War II (29 percent of the market in 1980), but still is confined largely to housing units marketed at the lowest end of the price spectrum (82 percent of all housing units sold for under $50,000 in 1983).

For several reasons the housing sector of the U.S. industry has almost no experience in international markets:

- The small size of most companies limits available funds to explore markets in other countries;
- Home building technology is based primarily on wood-frame construction, which is not the case in the rest of the world;
- Housing programs in most other countries are largely influenced by their governmental policies, and are not open to the speculative building characteristic of the United States.

The sectors termed "office and commercial," "institutional," "hotels and motels," and "all other private work" are influenced by the availability of a combination of land and financing packages. In recent years a large amount of investment capital from other countries has been placed into this sector. For example, the *Los Angeles Times* indicates that 75 percent of the large, downtown office buildings in Los Angeles are foreign owned or controlled, which is up from 25 percent just eight years ago. As will be discussed, such investment sometimes brings with it foreign constructors.

The heavy-construction sector ("industrial," "electric power," and "other utilities") generally involves the work of large firms, many of which participate in the international arena. Foreign heavy-construction firms, which tend to be large in size, are now looking to this area in the United States as a source of market growth.

The federal, state, and local government sectors generally attract firms that concentrate on government work because of the special marketing skills, and sometimes special political visibility, needed to gain work from governmental units. While government contracting requires open bidding, it is not always possible or desirable for many construction firms to bid on such work. Architectural and engineering firms do not bid on government work (although from time to time there is pressure from legislators to have them do so), but qualifying for consideration on government design awards takes a very different business strategy than getting design contracts in the private sector. In general, the markets for work with government clients have become increasingly price-competitive, reducing some firms' ability to invest in new technology.

TABLE 6 Major U.S. Contractors Working on a Global Basis (in $million), 1986

Company	Construction Contracts	
	Foreign	Total
The M. W. Kellogg Company, Houston, Texas	5,085.0	6,945.0
The Parsons Corporation, Pasadena, California	3,823.3	6,408.9
Bechtel Group, Inc., San Francisco, California	3,439.0	7,079.0
Brown and Root, Inc., Houston, Texas	1,818.3	3,540.6
Lummus Crest, Inc., Bloomfield, New Jersey	1,760.0	2,335.0
Foster Wheeler Corporation, Livingston, New Jersey	1,219.0	1,847.0
Fluor Daniel, Irvine, California	985.3	6,075.3
Santa Fe Braun, Inc., Alhambra, California	630.0	710.0
Stone and Webster Engineering Corporation, Boston, Massachusetts	428.0	1,625.6
Jacobs Engineering Group, Inc., Pasadena, California	275.5	982.3
Kaiser Engineers, Inc., Oakland, California	267.7	945.5
Dillingham Construction Corporation, Pleasanton, California	169.3	1,121.6
Fru-Con Corporation, Baldwin, Missouri	159.5	672.5
Kiewit Construction Group, Inc., Omaha, Nebraska	147.2	1,262.5

Note: Of the total global construction market of $73.9 billion (available for bids from outside of client country), 43 American firms obtained $22.6 billion (30.6 percent). The 14 firms shown in this list had more than 90 percent of the U.S. volume.

Market segmentation and the preponderance of small firms preclude much of the U.S. construction industry from international business. Of the top 400 U.S. contractors listed in *Engineering News Record* in 1987, 54 are involved significantly in international competition. The 14 largest firms account for more than 90 percent of U.S. construction work abroad (see Table 6).

Forty percent of the 500 largest U.S. design firms are involved in international work. The 22 firms listed in Table 7 were responsible for more than 85 percent of the work.

Construction Machinery

The United States has about 800 construction machinery producers, many of which export (or manufacture abroad) machinery to about 150 foreign countries. The primary markets are Canada, Saudi Arabia, Australia, and many nations in Western Europe. The large producers have quite extensive dealer networks around the world, both for sales and service.

TABLE 7 Principal U.S. Design Firms Practicing on a Global Basis, 1986

International Billings	Service[a]
$30 million or more	
Louis Berger International, Inc., East Orange, New Jersey	CE
Daniel, Mann, Johnson, and Mendenhall, Los Angeles, California	AE
DeLeuw, Cather and Company, Washington, D.C.	EA
Gibbs and Hill, Inc., New York, New York	EA
Harza Engineering Company, Chicago, Illinois	CE
Holmes and Narver, Inc., Orange, California	EA
Metcalf and Eddy, Inc., Wakefield, Massachusetts	EA
Morrison-Knudsen Engineers, Inc., San Francisco, California	CE
Under $30 million	
Black and Veatch, Kansas City, Missouri	EA
CRS Sirrine, Inc., Houston, Texas	AE
Camp Dresser and McKee, Inc, Boston, Massachusetts	CE
Dames and Moore, Los Angeles, California	CE
A. Epstein and Sons, Inc., Chicago, Illinois	EA
Gilbert Associates, Inc., Reading, Pennsylvania	EA
Frederick R. Harris Inc., New York, New York	EA
Lester B. Knight and Associates, Inc. Chicago, Illinois	AE
Charles T. Main Inc., Boston, Massachusetts	EA
Pacific Architects and Engineers, Inc., Los Angeles, California	EA
Parsons, Brinckerhoff, Inc., New York, New York	EA
Skidmore, Owings and Merrill, Chicago, Illinois	AE
Sverdrup Corporation, St. Louis, Missouri	EA
Williams Brothers Engineering Company, Tulsa, Oklahoma	CE

[a]AE = architect/engineer; EA = engineer/architect; and CE = consulting engineer.

Note: Of the total global volume of $3,543 million in design fees available to design firms from outside the client country, 49 American firms captured some $917 million (25.9 percent) of this total. The 22 firms shown on this list were responsible for more than 85 percent of the U.S. share.

The value of U.S. exports of construction equipment was at its peak of $6.3 billion in 1978 and has declined steadily to about $2 billion today. Similarly, U.S. employment in the equipment industry reached its peak in 1979 at about 175,000 workers and has declined by two-thirds.

Caterpillar Tractor Company of the United States is the world's largest construction machinery producer, with Komatsu, Ltd., of Japan following. There is at present substantial excess capacity in the world's construction equipment industry, and cost-reduction measures, more efficient and less costly manufacturing methods, and other similar measures are being undertaken by the producers. While price competition will probably remain the dominant factor in the industry, investments in research and development may yield future advances. For example, the development of more automated equipment extending the range of weather conditions under which construction is possible may be forthcoming.

FOREIGN FIRMS IN THE U.S. MARKET

The U.S. construction community faces a new challenge in terms of both cooperation and competition. With the general slowdown in other parts of the world, design firms and contractors from other countries see the very large American market as an attractive way to maintain or increase their business opportunities. As Case Study 2 illustrates, companies from Europe, Japan, and South Korea have been developing working arrangements in this country for some time.

In the five years from 1978 to 1982, the number of foreign design and construction firms entering the U.S. domestic market grew annually at rates of almost 8 percent and 13 percent, respectively (see Table 8). Revenue of foreign firms in the United States increased

TABLE 8 Foreign Design and Construction Firms in the United States

Category	Number of U.S. Affiliates		
	1978	1980	1983
Design and engineering services	40	53	58
Construction	45	70	82
	U.S. Income to Foreign Owned firms ($millions)		
	1978	1980	1983
Design and engineering affiliates	669	694	892
Construction affiliates			
European	1,142	3,896	5,394
Canadian	61	243	144
Japanese	24	50	81
Other	317	415	1,308
Construction total	1,544	4,604	6,927

Source: U.S. Commerce Department, Bureau of Economic Analysis.

Note: By 1985 the Japanese construction volume in the United States had increased to more than $1.5 billion, making Japan's penetration of the U.S. market the most dramatic.

during that same period at an annual rate of 35 percent. Japan's volume has shown stunning growth, reaching more than $1.5 billion by 1985. While total foreign work in the United States is only about 2 percent of the domestic market, it is concentrated in the large and technically complex areas of work that have been the mainstay of U.S. international business. Experts in the field find the situation alarming.

CASE STUDY 2: JAPAN'S OHBAYASHI GUMI: DOING CONSTRUCTION IN THE UNITED STATES FOR 20 YEARS

In the early 1920s, the California-based Fluor Corporation invited executives of a Japanese construction company to America to study advanced U.S. construction technology. Today, this company is back, bringing advanced Japanese construction technology with it. Over the past 20 years and more, Ohbayashi Corporation has built dams, tunnels, offices, and residential projects in the United States.

Founded in 1892 by Yoshigoro Ohbayashi, the company has been among the Big Five Japanese construction companies, which include Kajima, Taisei, Shimizu, and Takenaka Komuten. (Today, with Kumagai Gumi, they are the Big Six.)

Ohbayashi is among the world's most experienced dam builders. Its finished dams number in the sixties. It has been a leader in the development of roller-compacted concrete dams, as well as the use of deep concrete cut-off walls to control subsurface seepage.

Ohbayashi spends significant sums of money on research and development. It has one of the finest research facilities in Japan, the Ohbayashi Technical Research Institute, where the firm develops clean rooms for hospitals and semiconductor factories; superstrong concrete for nuclear reactors; concrete for use in underground continuous walls; computer software for complex engineering calculations, analyses, and simulations; polymers that prevent cave-ins; energy conservation systems; and other technologies. The firm has developed a dynamic suspension method that substantially mitigates damage to a building during an earthquake.

Ohbayashi has built a reputation for modifying existing technology to fit the job at hand. For example, it replaced the shield on its tunnel boring machine with a backhoe-like excavator on a major project in Phoenix, Arizona.

Ohbayashi's adaptation of the New Austrian Tunneling Method (NATM) improves on that technique. The NATM process uses rock bolts driven into the circumference of the tunnel to provide reinforcement. Concrete is then sprayed on the tunnel wall with an Ohbayashi-developed concrete distributor robot.

Ohbayashi did its first work outside Japan in Cambodia, building an agricultural center. Since then, it has done much work in Southeast Asia, including buildings, tunnels, and dams. In 1984, the company won a major contract from the People's Republic of China, for technical supervision of construction work on the Shanghai International Airport.

Ohbayashi came to the United States in 1966, opening an office in Honolulu, Hawaii, and registering to do construction there. In the same year, it began construction of the Surfrider Hotel in Waikiki as a construction manager. The hotel was finished in 1969. It also built the Princess Kaiulani Hotel in Honolulu, which was completed in 1970. Both hotels had Japanese owners, and the general contractors under construction management by Ohbayashi were Americans.

In 1972, a subsidiary company, Ohbayashi Hawaii Corporation, was established to engage in real estate development in Hawaii. Since 1972, this fully owned subsidiary has been developing real estate complexes throughout the islands. Also in 1972, Ohbayashi came to the West Coast and established its wholly owned Ohbayashi America Corporation (OAC), a general contractor, in Los Angeles. OAC is currently involved in a low-income housing complex for the Los Angeles City Redevelopment Authority and is developing a large-scale shopping center in the Little Tokyo area. OAC's major local affiliates are 2975 Wilshire Company, for office rental management, and James E. Robert, Inc., for condominium and apartment development in northern California.

In 1974, after two years in Los Angeles, OAC won a hotel construction contract, Kyoto Inn, located in San Francisco. The owner, Kintetsu, also owns a Japanese railroad company. Following the hotel project, OAC undertook banks, offices, restaurants, and housing contracts, mostly for Japanese clients.

In 1976, Ohbayashi participated in developing a large-scale residential complex near Seattle, Washington. This was a joint venture with Tokyo Corporation. The 11,000-acre site in Mill Creek includes a golf course, shopping center, and 3,200 housing units.

In 1979, Ohbayashi Corporation formed a joint venture with a local company to bid for a San Francisco sewage tunnel project, the first U.S. public work Ohbayashi was to undertake. Expertise in soft ground, using the earth pressure balance shield tunneling method, led to success in bidding on this project. The method cut costs substantially, compared to alternative methods. The owners were the city and county governments of San Francisco. Ohbayashi's San Francisco office became its headquarters for heavy construction in the United States and in 1981 the heavy division successfully bid the Strawberry Tunnel in Utah, a federally funded project.

In 1982 a New York City office opened, and in 1984 it won the construction contract for a 17-story building for a Chinese maritime company.

In 1985, Ohbayashi established a subsidiary in the Southeast named Citadel Corporation, headquartered in Atlanta. This was established from scratch, staffed and operated by Americans. An open-shop (i.e., nonunion) contractor, Citadel has been active in the region, completing five projects. The staff is American; the ownership is Japanese.

In 1986, Ohbayashi was selected as a construction manager for the big Toyota manufacturing plant in Kentucky. It is the largest project Ohbayashi has undertaken in the United States, entailing supervision of five American general contractors.

In 1987 the company beat U.S. competitors to win construction of tunnels for flood run-off in San Antonio, Texas.

Japanese personnel in U.S. Ohbayashi offices total 30 to 40 people. Some have studied engineering or management in this country.

Eiji Noma, general manager in New York City, who studied at the University of Chicago in the late 1960s, says it is more difficult now to get Japanese professionals to work in the United States. It is no longer their "hardship post" with perquisites and bonus pay, but rather an expensive place to live when paid in dollars, no better than living on yen at home. "That gap of income has narrowed, while hardships never lessened," says Noma. Still, every year four or five Ohbayashi people come to study in the United States, usually in the fields of engineering or management.

An Ohbayashi manager states the company's U.S. business objectives succinctly:

- *To satisfy traditional Japanese clients needing commercial or industrial buildings in the United States. To serve the needs of its Japanese clients is more important than to make money here.*
- *To compete and to do joint ventures with Americans for heavy construction work where Ohbayashi may have useful tunneling or dam-building expertise.*

In 1986 Ohbayashi contracted for $226 million worth of construction in the United States.

3
Competition in the Global Market

U.S. industry faces stiff competition in the international construction market. Foreign firms in many countries enjoy strong support of coordinated government policies that encourage export of services and enable these nations' firms to present a united front in competition.

COMMON CHARACTERISTICS

The committee's experience and review of limited available documentation reveal several common characteristics of these national policies. Outstanding features include central government leadership and strong financial support.

Many countries have a primary agency that takes responsibility for construction policy. In most cases, particularly in Japan and France, a government ministry at the cabinet level or a quasi-governmental entity deals with both domestic and international construction policy matters.

Collaboration within the full range of relevant organizations is apparent and includes leadership representing financial institutions, construction firms, research organizations, educational institutions, and development and export agencies in government. The composition of such policymaking groups reflects the comprehensiveness of the policy formulation process and the depth and breadth of the policy response.

Policies typically reflect a number of considerations. These include ramifications for using design services to gain follow-on construction, the relationship of construction to follow-on equipment sales, the relationship of construction to follow-on capital goods sales, and operations and maintenance aspects as well as replacement parts activity related to construction projects.

Policies also attempt to exploit national competitive advantage, that is, in which parts of the world the national industries have the most advantageous position and what elements of the competitive package are their strongest. In some cases, this has led to ranking technologies for further emphasis and investment and identifying target areas of focus for national investment.

Studies underlying policy often include some specific consideration of the nation's potential in the U.S. market. The U.S. market remains the largest stable and open construction market in the world (although the elimination of trade barriers in Europe's common market will create a combined market comparable in size to the United States). All of its Asian and European competitors have strategic programs for penetrating the U.S. market.

SPECIFIC CASES

The specific policies of several countries are instructive.

Great Britain

The government of Great Britain openly and clearly provides a number of mechanisms for supporting the efforts of British construction and engineering firms to obtain work on overseas projects. An Overseas Project Fund administered by the Department of Trade provides direct subsidies. In this arrangement, the government puts up a limited amount of financing to cover prebidding costs, with the subsidized firm required to return about 20 percent to the government if it is the successful bidder.

The government also engages in providing mixed credit,* using a "war chest" similar to that recently obtained by the U.S.

*Mixed credit is a means of reducing borrowing costs through provision of government-backed loans at concessionary rates together with commercial loans at market rates.

Export-Import Bank. This war chest is almost always used as an interest-balancing support and is obtained from British commercial banks. The funds so collected and utilized are administered through a governmental agency, the Overseas Development Administration, which, in many respects, is quite similar to the U.S. Agency for International Development. Requests for these monies from private sector companies are channeled through the Department of Trade.

The British Export Credit Guarantee Department (ECGD) is an institution which funds projects similarly to the U.S. Export-Import Bank and the Overseas Private Investment Corporation. Its resources are available for both engineers and contractors, and can be used to finance capital goods purchases.

The British construction and consulting industry prospered in the late 1960s and 1970s on projects in the Middle East, but with the severe diminution of that market, the industry has been forced to scramble for domestic work. Very few British firms have competed for World Bank projects because of the lengthy bidder list and because they are averse to the multilateral arrangements often required. Most British firms are privately owned, and several have made partnerships or other arrangements with U.S. counterparts, especially in the housing market. On the other hand, a number of the larger U.S. construction industry firms operate in the United Kingdom, especially in connection with North Sea oil field projects. Within the European community, the British have found it quite difficult to obtain projects, because of many administrative barriers, expectations of reciprocity, and no commonality of qualifications or standards. The British currently view the United States as their primary overseas target market.

The British have two organizations which, with governmental acquiescence and assistance, greatly assist their design and construction companies. The first is the British Consulting Bureau, headed by the Duke of Gloucester, which is active in developing such potential project areas as the People's Republic of China and Africa. Its member firms provide primarily engineering services, but also consultant services in health, agriculture, and various development disciplines. The second organization is the Export Group for the Construction Industries, whose purpose is to encourage others to use British construction companies on international projects. It, like the bureau, closely monitors overseas intelligence reports on potential projects, and receives strong support from the commercial sections of British embassies. The organizations act as a central intelligence point and disseminate the information to their members much more rapidly and accurately than is done by the U.S. Departments of Commerce and State.

France

The international activities of the French design and construction industry are backed by a French Ministry of Construction concerned with exports, construction economics, and global development. This ministry deals with such matters as cooperation and coordination among the construction firms in obtaining foreign projects and other export matters.

The ministry has also posted 40 persons in French embassies around the world, where they are considered to be investments in future projects of the host nations. The present international emphasis of the French construction industry is on urban systems, such as water, transportation, and nuclear power. There are also detailed analyses by the involved trade associations as to future market potential and concentration.

France has no specific government-sanctioned policies on international construction, but it does have an informal policy statement and understanding with industry. The French policies are reported to be based on an analysis that indicated the country receives a seven- to ten-fold return on each investment made in design and engineering projects in other countries. As in other countries, the French have found that its international markets peaked in 1980–1982 and have subsequently declined.

Italy

The first major international construction project by the Italians was a large dam in Zimbabwe, completed in 1956. By 1986 the Italian international construction volume had increased to a point where it stood third in the world, behind only the United States and Japan. Italy, in recent years, has concentrated on obtaining plant construction projects, rather than only civil works projects. Its international construction projects have included a $1.3 billion steelworks plant in the Soviet Union, a power distribution station in Saudi Arabia, a refinery in Greece, and the second Bosphorus Bridge in Turkey. In 1986 the Italians were working on 240 construction projects in 76 nations and 120 design contracts in 62 nations.

Three major groups, which include both design and construction firms, dominate the Italian construction sector. These are FIAT (through the Milan-based holding company Fiatimpresit); IRI (through the Rome-based holding company Italstat); and the League of Cooperatives. Among the top 50 Italian construction firms, 4 contractors belonging to the FIAT group (the largest private company in Italy) account for 15 percent of the total contracts; 6 firms belonging to IRI account for another 15 percent; and the 7 cooperative contractors have about 12 percent.

One of the reasons for the Italians' success is their ability to maintain a lasting presence in various nations, including Africa, Turkey, and Greece. Italian companies are fully competent to handle a wide range of rather specialized jobs, and close cooperation exists between the public and private sectors of the industry. The Italians have also come to realize the vital importance of "financial engineering" and to put forth proposals, both technical and financial, that are well suited to the needs and capabilities of the developing nations.

The Association of Italian Engineering and Techno-Economic Consulting Organizations (OICE, founded in 1966) represents its members to national and international client organizations. The efforts of OICE are directed at supplying clients with integrated technical and complex interdisciplinary solutions to plants, infrastructure, and engineering works in general. These services are not limited to technical and design services, but include organizational management and financial expertise, applied to both infrastructure and commercially oriented projects.

The Italians have long realized that to succeed in the international arena, a basic element is successful financial engineering. A major step was undertaken in 1977 when the Italian government organized a comprehensive and articulated program (the Ossola Law) to provide Italian exporters with the necessary financial support to compete successfully at the international level. However, Italian companies are still finding it difficult to compete against the mixed credit programs utilized by some other major nations. Thus, many Italian contractor companies use the intervention of specialized Italian investment banks with experience in the export credit field.

Sweden

In 1973 the Swedish government embarked on an ambitious housing program. Its goal of providing decent housing for every Swedish family called for the construction of 1 million houses per year for a 10-year period. This volume increased the capacity of the construction industry well beyond the "normal" market volume of previous years.

In 1983 a report of the Swedish Council for Building Research entitled *The Swedish Building Sector in 1990* set the foundation on which the next 10 years of Swedish construction activity will be based. The 1983 report concluded that a continued favorable expansion of building programs would be possible provided there is a substantial increase in expenditures on research, development, experimental construction, and demonstration activity. Even though construction has declined in recent years, it still represented 12.7 percent of Sweden's GDP, or $11.6 billion, in 1984.

A matter of concern in Sweden is the very low priority placed on research and development to retain and further develop technical competence. The Council for Building Research recommended that substantial increases in research and development expenditures by the government and the private sector over the next 10 years are essential.

A number of areas were singled out for attention in this research program:

- the development of building technology;
- satisfactory and economical property management;
- energy conservation;
- municipal planning;
- higher housing quality; and
- the role of the construction industry in the national economy.

The Swedish government supports technical research by the building industry as well as by technological universities.

From 1968 to 1979, the value of the Swedish export surplus of consulting services, construction abroad, and building materials increased almost tenfold. At the time of the 1983 report, the export of building materials and construction capability was of great importance to the Swedish economy, with about 100,000 people directly or indirectly involved in this market. The Swedish firms that compete in the international arena feel their competitiveness in foreign markets is often due to local ties and contacts in the host country (a perception shared by all major international construction firms). The Swedes are also convinced that companies exporting construction services must, in general, be sizable to be competitive.

In Sweden there is a centralized point of contact for the construction industry, the Ministry for Housing and Physical Planning, which is involved in both domestic and international matters and policies. One of its roles is to provide guarantees for international construction activities in order to help Swedish firms compete with other nations.

Japan

The design and construction industry of Japan is considered a unique phenomenon in both its overseas operations and in its domestic practices. For the past several years Japan has been of major concern to the nations with which it competes in international markets. This perception, however, is probably distorted by the huge export success Japan has had in such manufactured items as automobiles and electronics goods. There has been an assumption that the same phenomenon was, or could be, occurring in the construction industry. However, while Great Britain and West Germany each have over 8 percent of the total international design market, Japan's share in 1986 was only 6 percent, a nevertheless admirable figure in view of the nation's relative size. The United States, with 30 percent of the international construction market, competes as much with Italy, France, or Britain as with Japan (see Tables 3 and 4).

Japanese domestic policies on construction have been the source of frustration and misunderstanding on the part of those nations who wish to work in the Japanese market. The Kansai Airport project has been a recent and large symbol of this frustration for the United States. Despite apparent concessions, there can be but little doubt that the Japanese government is determined to protect a major share of such large projects for Japanese constructors.

In June 1987, the Economic Council of Japan issued a detailed set of policy recommendations for Japan's Economic Structural Adjustments. One of the recurring themes is a concern that the economic growth of Japan be truly reflected in the quality of life for its citizens. For example, the report indicates that the present state of the nation's infrastructure is considerably below that which the overall GNP would indicate it could be. There is, therefore, a potential domestic market of rather sizable dimensions for the Japanese construction industry. The council also recommends positive efforts to ensure that foreign companies can do business in the Japanese construction market, and expansion of the General Agreement on

Tariffs and Trade (GATT) framework to include design and construction services.

As in the United States, Japan has a handful of construction companies that dominate both the domestic and international markets and literally thousands of mid-size and smaller firms whose market is strictly domestic. The Big Six construction companies increased their share of the international market from about 1980 to 1985, and then their share began to decrease significantly. While the worldwide shrinkage of international projects was pivotal in the decrease, another factor was a self-imposed retrenchment. A report entitled *Overseas Construction Basic Issues: Investigation Committee,* sponsored in 1982 by the Japanese Ministry of Construction, emphasized that "there are many problems related to the short history of our overseas construction activities. Further development is expected to yield a genuine service export industry. However, the road is not necessarily smooth." Although this report on overseas construction was compiled in 1982, it remains the main guidance for the industry as a whole. There has not been a reason to revise or update it in the ensuing five years, according to the Ministry of Construction representative at the Embassy of Japan in Washington. Even though the experience of the Big Six contractors with U.S. offices has been that profits are poor to nonexistent, the companies do not dare, as yet, abandon the U.S. construction marketplace.

India

In 1986 the Indian government set aside $1 billion for a three-year period to boost its engineering sector, funded by a combination of World Bank loans matched by India's contributions from both the government and private sectors. With the second largest population in the world, there is potentially an enormous backlog of infrastructure work required within the country. India's design and construction firms, however, are more interested in working on projects outside the country, acting as subcontractors or joint venture partners with firms from the larger developed nations. This emphasis stems from two desires: technology transfer to the Indian firms involved, and an increase in foreign exchange earnings.

India has no central authority for construction and engineering, but these sectors are nominally under the purview of the Ministry of Housing and Public Works. India has a number of engineering and construction councils, most of which are private, that actively seek projects both within India and outside the country for their member companies.

The Soviet Union

The industrialization of the Soviet economy since the 1930s has given designers and builders of plants and large civil engineering projects the credentials required to work in the international arena. Most of this work is in the less-developed countries, especially those nations with large public sectors and socialist forms of government.

All design and construction activities of the USSR are organized within the mammoth agency known as Gosstroy. To export these capabilities the Soviets have formed about a dozen foreign trade organizations (FTOs) that are, in reality, large contractor organizations with formidable capabilities. Although generally confining themselves to proven Soviet technologies, the FTOs, on occasion, will design new plants, equipment, and infrastructure for their clients. For fundamentally simple phases of a given project, the FTO will usually depend on the local contracting abilities of the client country for basic construction. However, for heavy equipment and other more complex phases, the FTOs depend on their own sources of supplies, supplemented surprisingly and quite frequently with Western equipment and material.

Since all foreign projects are viewed as ventures of great prestige to the USSR, only the very best engineers and technicians are sent abroad. Although complaints are often voiced in the Soviet press concerning the calibre or slowness of domestic projects, these complaints are seldom heard on foreign ventures, which are turnkey type projects, with the project management subcontracted to Austrian and Finnish companies. Of the 65 FTOs in the USSR, the dozen that are allowed to engage in foreign projects have been licensed to form joint ventures with Western firms and to purchase supplies, technology, and equipment from Western suppliers.

Each FTO that engages in overseas projects has one basic specialty, with a number of other capabilities. These specialties include an FTO that is one of the world's largest suppliers of power-generating and transmission equipment, one that has built more than 600 industrial plants and communications facilities, one specializing in infrastructure projects, and one that is expert in iron- and steel-making equipment.

The magnitude of foreign activity may be judged by the fact that the USSR has signed agreements with 83 nations for economic and technical cooperation. A total of 3,054 international projects were completed between the end of World War II and 1986. These projects include 1,769 industrial enterprises and power-generating plants, and 329 agricultural projects.

U.S. RESPONSE TO COMPETITION

External economic forces have had substantial influence on how U.S. construction has responded to international competition. The high value of the U.S. dollar in international exchange has until recently had particularly strong impact on this response. Abroad, and at home, U.S. firms have appeared relatively more expensive than their foreign competition.

The strong U.S. dollar between 1980 and 1985 served as a magnet for imported goods and investment. A series of major tax cuts and increases in government spending during this period fueled a strong recovery in the United States while other industrialized nations consistently pursued slow growth policies. U.S. industry was placed at greater disadvantage in both domestic and international markets, with the result being stagnant exports and a rapid growth in import penetration of the U.S. markets.

However, as economist Robert J. Samuelson wrote in the January 26, 1987, issue of *Newsweek:*

> Real changes underlie our competitiveness anxiety. The United States no longer enjoys unchallenged superiority in trade and technology. Some of our supremacy was artificial: World War II destroyed our most potent commercial rivals. Europe's reconstruction restored this competition. The spread of technology, modern education and multinational companies to Japan and the developing world created new competitors. Reversing these trends is impossible. A competitive vision that reinstates the United States sitting astride global markets is pure nostalgia.

Nevertheless, the examples reviewed here illustrate that the United States has been slower than many of its competitors to develop national trade and economic policies in support of international engineering and construction. While industry and trade groups have been vocal in reporting the practices they face in the global market, the industry lacks both central representation in national policy discussions and the means to pull together diverse public and private interests to present a unified competitive front.

In addition, there are specific problems. Foreign policy considerations can make U.S. firms unacceptable in a country after substantial investments have been made in market development. The U.S. Foreign Corrupt Practices Act (FCPA), enacted by Congress for important ethical reasons, hampers the ability of U.S. firms to conform to local business and cultural standards. Competition from other countries not subject to such regulation can put American firms at a disadvantage in business negotiations.

Other more specific disincentives are found in U.S. policy:

- Income tax requirements for U.S. citizens working abroad impose a greater burden than those of other countries, making it more costly to provide incentives needed to attract high-quality personnel to foreign assignments.
- Double taxation occurs on design work performed in the United States for overseas projects, because foreign corporate taxes on imported engineering services may not be deducted from U.S. earnings.
- U.S. antiboycott laws that conflict with the boycott laws of other countries restrict opportunities open to U.S. firms.

The previously described activities of the U.S. Trade and Development Program, Export-Import Bank, and Overseas Private Investment Corporation provide valuable but severely limited assistance to U.S. design and construction firms seeking to provide competitive financing for projects. The increasing importance of finance has encouraged U.S. firms such as Bechtel, Fluor, and Kellogg to form consortia with British, French, German, and Japanese companies:

- Bechtel associated with American, French, and Japanese suppliers and export finance sources for the $450 million Rio Zulia to Covenas pipeline and associated facilities constructed for Ecopetrol and Occidental in Colombia.
- The Fluor Company built a pipeline for the Petroleum Authority in Thailand (PTT) as part of a consortium that included the four largest steel producers in Japan, with funding provided by the Bank of Tokyo.
- The Kellogg Company developed a cooperative agreement with the West German firm Thyssen to undertake a $1 billion aromatics project in Indonesia.

U.S. companies bring specialized technological skills and managerial expertise to these consortia, while their partners provide the financial support through their own government agencies, which can provide money and guarantees to support the export of services, materials, and equipment. Indeed, these specialized technical skills have been the source of U.S. competitive advantage in the past, although this advantage is not exclusive, as the work of Shimizu with IBM illustrates (Case Study 3).

American companies still are generally given high marks for their abilities in design, engineering, project management, and the operation and maintenance procedures for facilities. U.S. firms still lead the world in the design of process plants for the petroleum and petrochemical industry, as well as the technologies of chemical plants and power stations. The nation leads in the use of computer-aided design and drafting techniques, and the use of computer-based tools for construction management, scheduling, and inventory controls. However, the present shortage of large projects around the world reduces the advantage of this management know-how as price becomes increasingly decisive in client decisions. Of greater long-term importance is the concern of industry leaders that other countries are catching up with and passing the United States.

CASE STUDY 3: SHIMIZU MEETS IBM'S NEEDS

In the spring of 1986, IBM faced its greatest construction challenge in more than a decade. Its prime semiconductor development and manufacturing facility in East Fishkill, New York, needed a new technology center of approximately 300,000 square feet. And occupancy was required in less than two years.

The new Advanced Semiconductor Technology Center (ASTC) was described by operating management as an important milestone, playing a role in the future of IBM and its ability to remain competitive in the development and manufacture of advanced semiconductor products. The statement of requirements called for levels of environmental purity and vibration resistance never before achieved within the company. IBM management wanted the new building to be the best in the world.

Design and construction of the new building would be the responsibility of the Real Estate and Construction Division (RECD), which began a search for an outstanding semiconductor facility design firm. In view of the considerable accomplishments of Japanese companies in the design and construction of semiconductor facilities—including an IBM plant in Yasu, Japan—RECD considered two Japanese firms, Shimizu and Ohbayashi. RECD management also considered several U.S. design and engineering firms for the project.

IBM recognized that considerable development studies would be required during the design stage and that close coordination would be required between the design and construction people.

RECD representatives visited Shimizu facilities in Japan in the late spring of 1986; they came away favorably impressed with what they had seen and learned.

Shimizu is over 180 years old and one of the five largest design and construction companies in Japan with annual sales of over $6 billion. The company has offices in the United States, including a New York City location. Most of the work Shimizu had done here had been for Japanese companies with U.S. operations.

Shimizu has an annual research and development (R&D) budget of $60 million, which is 1 percent of annual company sales. This is typical of major Japanese design and construction firms. In contrast, RECD found only limited research at either design or construction firms in the United States. At Shimizu, 630 people are engaged in R&D work on systems development, product technology, infra-

structure engineering, intelligent buildings, construction automation, robots, and clean room design.

Shimizu's work in clean room design and vibration prevention was particularly noteworthy, and applicable to semiconductor facilities. Shimizu had achieved class 1 capability for particles of 0.5 to 0.3 microns in size. A considerable amount of the R&D activity was in testing filters. The firm had a large vibration table to conduct seismic tests on structures from which it developed state-of-the-art designs.

The RECD team also reviewed Shimizu management systems for planning, cost estimating, scheduling, and project control, which are very similar to those used by U.S. construction companies. A construction job was also observed. Shimizu is basically a construction company. It will construct a design prepared by another firm, but would not produce a design to be constructed by another company. RECD also reviewed the often difficult working conditions at East Fishkill with Shimizu.

RECD then recommended to senior IBM management that Shimizu be hired to design and construct the ASTC project. Management agreed, and in July 1986 Shimizu began working with an RECD engineering team. The goal was to develop a design concept based on the IBM requirements and design criteria. Shimizu established a base of operations across the street from RECD's headquarters in White Plains, New York.

Overall, the working relationship between the IBM and Shimizu teams went well. There were language and cultural differences to overcome; however, as the participants worked together, sound mutual respect was developed. IBM was very impressed with the skill and dedication of the Shimizu designers. They often worked "round the clock" to answer questions (an expedient in view of the time difference between the United States and Japan). Also, work in support of the U.S. team was done at Shimizu's R&D facilities in Japan. The turnaround time on most of this work was excellent. In the area of administration, the design contract took much longer to negotiate than a like contract with a U.S. design firm because of the unfamiliarity of the people with each other. Also, some difficulties with Shimizu's billings were experienced by IBM because of the absence of supporting detail.

Shimizu and IBM spent considerable effort on cost estimates, the mutual definition and understanding of costs, and the negotiation of the cost of work. Shimizu's initial cost of construction work was *about 10 percent more than the IBM budget. By working closely together, IBM and Shimizu came to a project cost agreement in the summer of 1986, which was reiterated in November 1986.*

In the early stages of design, Shimizu hired a U.S. architectural engineering firm, Giffels Associates, Inc., of Southfield, Michigan to share the design work. Giffels was chosen in part by Shimizu because of knowledge of conditions, local codes, and working practices at East Fishkill, where during the prior five years the U.S. firm had designed many facilities. Although Shimizu provided the design direction, a considerable amount of the work was done by Giffels. Shimizu also set up a liaison team in Giffels offices.

The Shimizu design had a strong bias toward initial cost effectiveness in contrast with future lower maintenance costs. Life-cycle cost appeared to be a lesser consideration. Overall, the design work proceeded well although the working drawings fell behind schedule.

In November 1986, Shimizu hired Huber Hunt & Nichols (HHN), Inc., Indianapolis, as general contractor on the construction of the building. Shimizu chose HHN because of the successful work the firms had done together on other U.S. projects, plus HHN's familiarity with working in the East Fishkill area. Although as of that date IBM and Shimizu had a contract for the design work only, it was the intent of both parties that Shimizu would manage the construction phase.

Sufficient design had been completed to begin the construction work of grading, footings, and foundations. In December 1986, a groundbreaking ceremony was held.

The first schedule difficulties arose in January 1987, when 50 percent of the working drawings were due for bidding purposes; only 15 percent of the drawings were complete. Nevertheless, the subcontractor bidding process began. On February 2, Shimizu reiterated the budget cost which had been agreed to in 1986. On February 13, Shimizu and IBM officials met at RECD headquarters in Stamford, Connecticut. Shimizu said that the project's cost of construction had increased—by nearly 40 percent—over the cost of record. They could not explain the cost increase except to state that it was based on inputs of the general contractor and subcontractors. The initial occupancy schedule date had also slipped. Inasmuch as IBM had not changed project requirements, the new Shimizu cost was rejected by IBM.

A month later, IBM, with inputs from Shimizu and others, was able to define the reasons for the cost increase, which can be summarized as follows:

- *Shimizu seemed to experience difficulties in working with the subcontractors in the East Fishkill area. In Japan, much work is done on not much more than a handshake between the parties. Here, the subcontractors appeared apprehensive about working for a foreign contractor. The language and custom differences, which were overcome by Shimizu, IBM, and Giffels during the design work, could not be surmounted during the comparatively short bidding cycle. The role of HHN in the bidding was less than one would expect of a general contractor.*
- *Shimizu seemed to have limited confidence in U.S. specialty products, manufacturers, and supplies. Their designers wished to specify many items from Japanese suppliers with whom they had extensive experience.*
- *Shimizu seemed to expect that U.S. client companies such as IBM would approve the budget cost increase, trusting Shimizu's efforts as the best possible.*

It should be noted that Shimizu accepted the East Fishkill area labor practices, productivity levels, and so on as a given, whereas IBM believed a fresh approach based on the Japanese model could yield some improvements here, as has been the case in the automotive industry. IBM was also disappointed that Shimizu's guarantees went no further than HHN's guarantees, which in turn were based solely on the inputs of the subcontractors.

Intense negotiations with Shimizu failed to result in a cost decrease. Therefore, IBM requested that Shimizu complete the design and act as a consultant during construction, but not as the construction manager.

Shimizu continued with design completion while IBM began intense negotiations with U.S. contractors to do the construction. These negotiations were successful, and the project was awarded to Walsh Construction Company of Trumbull, Connecticut. The project cost is now within the IBM budget, albeit at a higher number than the original Shimizu contract and with less contingency. The project will be undertaken on a phased basis in view of the schedule delays that were experienced. (Shimizu does not recommend this approach as it is more difficult to guarantee project quality.)

Initially, Shimizu was reluctant to act as a consultant because of its corporate policy not to contract for design work without actually managing the construction. However, ultimately IBM and Shimizu signed a consulting agreement. IBM's practice is to retain the design firm to support the construction.

The value of an IBM/Walsh/Shimizu relationship during construction is that the involvement of Shimizu will better ensure that the project is built per the plans and specifications. Shimizu will gain valuable experience in the U.S. market, and Walsh will have the benefit of a quality-oriented associate with an intimate knowledge of the design.

4
Research and Development in Construction

Research and development (R&D) in construction includes a broad range of activities directed toward improving quality, productivity, and efficiency of the materials, equipment, labor, and management of construction. The value of R&D activities is well accepted as means for improving productivity and generating new ideas in electronics, telecommunications, genetic engineering, and other technical fields. The linkages between construction research and application, however, have been more difficult to document, despite advances made during the twentieth century in new equipment and materials, largely because of the great number of mostly small-scale builders and equipment and materials producers. For this same reason, the construction industry has greater difficulty mobilizing resources needed to support substantial research programs.

As a result, the committee observed several troubling trends:

- Other countries appear to be putting more effort than the United States into construction R&D;
- Other countries are working hard to improve the "hardware" of construction by improving construction methods and developing technology for automation (including robotics);
- A more innovative environment exists in most foreign firms because R&D has been integrated into overall operations;

- Other countries are willing to back longer-range research efforts through the slow but methodical methods needed;
- R&D in other countries tends to be proprietary to the company sponsoring it, leading to some duplication but increasing commercial rewards for success;
- Vertical integration within large foreign construction firms has made easier the utilization of research results by the operating units of their companies;
- There is less emphasis on research related to the "management" of construction by firms in other countries, since they tend to acquire these technologies through joint ventures with American firms or by sending their young professionals to U.S. universities for training.

U.S. CONSTRUCTION RESEARCH AND DEVELOPMENT

Accurate appraisals of R&D investments in the U.S. design and construction industries are stubbornly elusive. Available statistics are scarce and often recorded in a manner that can be misleading. In another study* done by the Building Research Board the following observations were made on R&D expenditures in the U.S. design and construction industries:

- Construction contractors (both general and specialty)—$54 million
- Manufacturers of construction materials and equipment—$838 million
- Federal agencies (both consumers and nonconsumers)—$200 million
- All other sectors (based on estimate)—$111 million
- Total annual construction-related R&D—$1,223 million

Based on a total volume of construction of some $312 billion in 1984, these estimates represent about 0.4 percent of sales invested in R&D, far less than other mature industries such as appliances at 1.4 percent, automobiles at 1.7 percent, or textiles at 0.8 percent. (This expenditure level is also well below Japanese construction R&D expenditure rates.) U.S. contractors, architects, and engineers invest less than 0.05 percent in R&D as a group, a fraction of the amount they spend on liability insurance alone.

---

* *Construction Productivity,* National Academy Press, Washington, D.C., 1986.

Both lack of resources and competing priorities are factors in this low level of R&D expenditure. Faced with intense price competition, many designers and constructors find it difficult to appropriate substantial resources for R&D. Tax regulations that may require capitalization of R&D expenditures increase the demands R&D would make on current cash flows. The natural aversion to risk of many businessmen makes R&D spending that may yield no immediate commercial benefit—more difficult to justify even when business is good, and easy to cut when times are bad. Not one of the many medium and small firms can afford a meaningful research program, and there are few mechanisms to facilitate joint funding of research that will yield distinct benefits to the participating firms.

What the optimum level of U.S. construction R&D spending ought to be is a complex question for which the committee found no ready answer. Observation of U.S. performance in introducing technological innovation and an eroding competitive position make it apparent that the level of spending—viewed either as an investment for increased productivity or as an indication of openness to new ideas—is too low.

Direct government involvement in construction research is limited but significant:

- The National Science Foundation (NSF) has been a principal source of support for university-based research activities for the U.S. design and construction industries. Through the NSF, National Engineering Research Centers are being established, such as the Center for Advanced Technology for Large Structural Systems (ATLSS) at Lehigh University. In addition to NSF funds of $10.4 million over a five-year period, other state-related institutions and the private sector are providing matching funds. The major goal of the ATLSS center is to do research and develop technology benefitting U.S. structures-related industries in design, fabrication, and construction, and inspection and protection of structures in service.
- The federal government laboratories such as the Army's Construction Engineering Research Laboratory (CERL), the Navy's Port Hueneme Civil Engineering Laboratory, the Tyndall Air Force Engineering and Services Research Center, and the National Bureau of Standards' Centers for Building Technology and Fire Research conduct research on a diverse range of topics with military and civil applications.

- Grants from the Army Corps of Engineers have produced major new research programs at the Massachusetts Institute of Technology and the University of Illinois.

The Construction Industry Institute at the University of Texas at Austin is an outstanding example of research without direct government support. More than 65 organizations representing owners, contractors, and 25 academic institutions have combined their resources to tackle advanced construction research. The institute then represents an important model for broader public-private partnership in construction research.

OTHER EFFORTS NEEDED

An examination of research ideas for addressing societal needs, undertaken by the Technical Council on Research of the American Society of Civil Engineers in 1979, indicates a long list of research suggestions, most oriented toward improving the methodology of engineering. The list includes a large number of projects related to improving methodology, many of which could be valuable in the international arena.*

The architectural research community is based almost exclusively in universities, so that the potential exists for linking such research to teaching programs. The civil engineering research community is also largely based in universities, but there is some mechanical, electrical, or electronic research of direct relevance to the construction sectors being done by these other departments. To a limited extent both architectural and civil engineering research institutions do projects related to mechanical and electrical systems. Most research institutions have projects tied to computer-based design and engineering, but more work is needed, particularly to bring new results into practice, through teaching and professional outreach programs.

While spending on research often exceeds U.S. rates, the work going on in construction sector research programs in other countries tends to mirror programs in U.S. universities and government laboratories, with three major exceptions:

---

* *Addressing Societal Needs of the 1980's Through Civil Engineering Research,* The American Society of Civil Engineers, New York, New York, 1979.

- The work supported by the Swedish government on behalf of the building industry tends to be much more people-oriented, describing user requirements and how these requirements should be accommodated in design. However, there does not appear to be any better match between the research programs and the teaching programs in the universities than in the United States.
- The Soviet Union has six major research units within its construction agency Gosstroy. Five of these units do traditional science and engineering research of the type done in government building laboratories around the world, but one research unit concentrates on "cybernetics." Not much is known about the work of this unit, but it potentially could represent an interesting area for collaboration.
- With their government's strong encouragement, the six large, integrated Japanese construction companies all support research by internal units. These programs include hundreds of people, excellent facilities, and a broad spectrum of subjects (see box).

This committee has not undertaken to recommend a complete agenda for research in construction and design, and planning of such an agenda by a single centralized body would in any case be unproductive. However, committee members feel that certain types of research are clearly needed, such as these two examples:

1. The general subject of "diagnostics" is talked about within the architectural research community as an area for methodological improvement. Work on this subject could be greatly enhanced if university researchers and practicing architects worked in parallel with firms that are in the business of designing and marketing diagnostic instruments. A program that provides special funds to research units (as contrasted with individuals) within universities that had already obtained an agreement for matching funds from instrument companies would encourage vertical integration between the architectural sector and the equipment-producing sector.
2. The development of safety methods for structures during the construction phase could benefit from case studies. For example, the NBS Center for Building Technology has just completed a study of the collapse of L'Ambiance Plaza in Bridgeport, Connecticut, a building which was being constructed using the lift-slab method. This collapse could serve as a case study for a structural engineering faculty to develop a continuing education course for engineers in practice, thus providing a link among a federal laboratory, university research, and professionals. While this subject is unique and timely,

THE JAPANESE CONSTRUCTION INDUSTRY AND R&D

Japan has established a Ministry of Construction responsible for setting national policies on behalf of the construction industry. One of its major policy decisions was to encourage private firms to establish research and development (R&D) capability. As a result more than 20 of the largest firms in Japan now invest 1 percent of their sales in R&D, and in these construction firms R&D has become a way of life. Each has established well-equipped, campus-like research centers, and research is integrated throughout their operating divisions.

The government of Japan provides a tax deduction for R&D of up to 1 percent of revenues, sometimes provides loans, and sometimes sponsors research projects directly. University-based research is relatively limited by U.S. standards, but the government funds and operates a Building Research Institute and a Public Works Research Institute.

The large private construction firms in Japan invest a small portion of their research funds in economic and marketing studies of what they should be designing and building, but much more goes into such technical subjects as new materials and design ideas. Their laboratories are furnished with the latest equipment:

- Shake tables for earthquake simulation;
- Wind tunnels for analysis of structural designs;
- Environmental chambers for evaluating performance of mechanical equipment;
- Sound chambers (both quiet and noisy condition chambers);
- Structural testing devices;
- Fire testing equipment;
- Materials and chemical testing laboratories;
- Clean rooms for more high-technology work;
- Hydraulic and geotechnical laboratories for civil works projects; and
- Outdoor testing yards for long-term analysis of weathering.

In addition Japanese companies do work to improve design and construction processes through applications of computer-aided design and engineering systems, new methods such as slurry walls in foundation construction, and construction automation and robotics. They are working in other fields as well:

- Biotechnology to improve the quality of lake and river water and develop a new water treatment processing system for sewage and industrial waste;
- Mechatronics, including robotization, teleoperation technology, automatic controls, and construction work control systems;
- Application of fifth-generation computers including computer-aided planning, design and construction, maintenance, and engineering;

- New energy sources, including coal gasification, fuel cells, solar cells, solar thermal systems, coal liquefaction, new batteries for energy storage, and hydrogen energy; and
- New primary materials, for example, the addition of metals, plastics, ceramics, and electronics to conventional materials, such as soils, rocks, cement, asphalt, and steel.

It seems likely, at the moment, that the people of the United States will benefit more from the Japanese strategy (by importing improved infrastructure in the future) than from existing infrastructure research in the United States.

the concept is to have this work serve as a model for similar projects on a range of structural safety problems and solutions.

As will be discussed further in Chapter 6, the development of advanced concepts for infrastructure poses an international challenge of enormous proportions. The present practice of dealing with urban transportation, water and energy supplies, waste management, and communications is based on inventions developed nearly a century ago. In the largest cities of the world these old inventions are clearly not well suited to dealing with present problems, and in the small communities of the developing world there has always been a kind of hand-me-down, makeshift quality to the nature of infrastructure investments.

New technology for infrastructure could possibly help the United States avoid the endless cutting and patching of our 100-year-old systems, and could also provide whole new market opportunities in the international sphere. There should be special programs to concentrate on infrastructure development within the university research community. These programs should encourage university units that are skilled in the areas of the "emerging technologies" to explore ways of creating new or higher-performing systems for infrastructure. Technologies such as new ceramics, advanced microelectronics, biotechnology, and genetic engineering should be incorporated into joint programs with the architectural and civil engineering faculties, and especially to provide graduate students from these technological areas the opportunities to work on infrastructure. In such programs universities could associate with trade and professional groups, such as the American Public Works Association, to introduce engineers in practice to new technologies and their capability.

The committee recognizes that some engineering schools can best be encouraged to expend research and teaching in construction by evidence of employment interest for their graduates. Programs may be needed to link employers with graduate programs in construction by having the university offer special graduate programs for mature employees of professional firms.

As the Japanese model illustrates, university-based activity is not the only way that construction R&D can be accomplished, but in the United States, academic institutions have become the primary centers of research. This pattern is unlikely to change in the foreseeable future, nor is it clear that it ought to change. What is clear to the committee, however, is that better mechanisms for linking research to construction practice are needed.

There is a need as well to increase the speed with which ideas from one field of research are tested for their value in other fields, and with which ideas of value enter practice. The case of the Bell Laboratories (Case Study 4), drawn from an industrial situation very different from construction, is nevertheless instructive because of their great success in linking research to the market. In construction, where the market is distributed among so many suppliers and buyers, projects built with federal government funds can be used to demonstrate new technology. A good example is the introduction to U.S. transit construction of precast concrete segmental tunnel liners (see Chapter 6).

The U.S. Department of Commerce has noted, "Over the next twenty years it is totally reasonable to expect that we will see widespread application of the following technologies: advanced materials, microelectronics, automation, biotechnology, computing, membrane technology, superconductivity, and lasers."* Today and in the near future many other new technologies may be added to the list. Mechanisms are needed to expose these new technologies and construction to one another, and to produce design and construction professionals competent to make the connections required for innovation. Besides institutional research, there must be training and education.

**Effects of Structural Change in the U.S. Economy on the Use of Public Works Services*, U.S. Department of Commerce, Washington, D.C., 1987.

CASE STUDY 4: THE BELL TELEPHONE LABORATORIES

The invention of the telephone is perhaps the single best modern example of how new technology can alter building and infrastructure. The Bell Telephone Laboratories have for more than 60 years been one of the leading U.S. centers of research and innovations that have changed how to design and build individual structures and cities, as well as the more basic structure of the economy and society.

The committee recognizes that the Bell Labs are a product of a private sector monopoly company that had vertical integration and an ability to make effective decisions about resource allocation and management strategy, with greater ease than is the case in U.S. design and construction. Nevertheless, many characteristics of the Bell Labs can serve as a useful model for institutional arrangements needed to strengthen U.S. building research. It is instructive to look at the history and accomplishments of this organization:

> The invention of the telephone was not inspired by a pre-existent popular demand. Rather, it came about largely through the ingenuity and vision of one man—Alexander Graham Bell. His belief that there was a great potential need for two-way voice communication over a distance, a need of which few men had been conscious, was confirmed by its immediate success and spectacular growth in spite of early technical limitations.
>
> By the end of the first fifty years a great new industry had been developed. There were nearly seventeen million telephones in the United States, almost twelve million of them in the Bell System. And in perhaps no other field had the force of scientific research in support of engineering development been so effectively demonstrated.*

As the AT&T Company Annual report for 1913 said:

> At the beginning of the telephone industry there was no art of electrical engineering nor was there any school or university conferring the degree of electrical engineer. Notwithstanding this the general engineering staff was soon organized, calling to their aid some of the most distinguished professors of science in our universities.
>
> As problems became more formidable and increased in number and complexity, the engineering and scientific staff was increased in size and in its specialization so that we now (1913) have working at headquarters on the problems of the associated companies some 550 engineers and scientists carefully selected with due regard to the practical as well as the scientific nature of the problems encountered.

**A History of Engineering and Science in the Bell System,* Bell Laboratories, Murray Hill, New Jersey, 1975.

> It can be said that this company has created the entire art of telephony and that almost without exception none of the important contributions to the art has been made by any government telephone administration or by any other telephone company either in this country or abroad.
>
> By 1924 the technical programs of the Bell System had so grown in range and intensity, and in number of personnel, as to suggest formation of a single new organization to handle most or all of these activities. Such an organization was formed on December 27, 1924, and started operations on January 1, 1925, under the name of Bell Telephone Laboratories, Incorporated. This corporation had a dual responsibility—to the AT&T Company for fundamental researches and to the Western Electric Company for the embodiment of the results of these researches in designs suitable for manufacture. At the date of incorporation, the personnel numbered approximately 3,600, of whom about 2,000 were members of the technical staff, made up of engineers, physicists, chemists, metallurgists and experts in various fields of technical endeavor. . . .
>
> Technological innovation had formed the indispensable core for telephony's growth up to 1925, but was even more significant to the future because so much of it was fundamental: the way was being prepared for more powerful systems yet to come, which would be essential to the enormous expansion felt to be lying ahead. Perhaps more significantly, the application of scientific methods to solving the "system" problems of telephony set a pattern that influenced industrial research and development by demonstrating the power of these methods and developing techniques of management that encouraged their use.
>
> Backing up the work on systems, which had laid the groundwork for so much that was yet needed, were the successful management techniques which had been developed for conducting and applying research, the means for closely controlling the quality of manufactured product, and a type of organization providing close integration of the user, technical developer, and manufacturer.

The Bell Labs have produced the transistor, the laser, the solar cell, and the first communications satellite, as well as sound motion pictures, the science of radio astronomy, and crucial evidence for the theory that a Big Bang created the universe. While they are a private laboratory (in the distinction made in the United States between government and private research work), their financial support was largely generated from a kind of tax on every telephone in the United States (before the breakup of AT&T in 1984), which in turn was allowed by their rate examiners (a public institution designed to monitor a monopoly utility). The potential for direct emulation by a government/industry research center is limited, therefore, but the operating principle of striving for increased systems performance by teams of scientists, engineers, manufacturers, and systems operators is a good one.

> Today's telephone caller uses components Bell never dreamed of, today's driver depends on systems Daimler and Benz never thought of, and today's homeowner switches on a power and light system that Edison never envisioned. These discoveries have long since been embedded in mammoth networks of technology that no single individual invented.
>
> Technological systems evolve through relatively small steps marked by the occasional stubborn obstacle and by countless breakthroughs. Often the breakthroughs are labeled inventions and patented, but more often they are social innovations made by persons soon forgotten. In the early days of a system such as electric light and power, inventors played the prominent role. Then as the system matured and expanded to urban and regional networks, others came to the fore. Electric light and power systems today are not just scaled-up versions of the Pearl Street station that Edison introduced in New York City in 1882. By the turn of the century, for example, it was the utility manager, not the inventor or engineer, who played the major role in extending round-the-clock service to many different kinds of customers—to the night shift chemical plant as well as the rush-hour electric streetcars.*

*Thomas P. Hughes, The inventive continuum, *Science 84,* November 1984.

5
Education and Training

Although the U.S. system of professional education continues to produce highly qualified engineers, architects, and construction managers, and to attract students from countries around the world, the committee nevertheless feels that change is needed. Experience in the international construction market shows clearly that young professionals need strength in four key areas to meet the challenges of global competition:

- A strong technical base;
- A clear understanding of design;
- An understanding of the intimate connection between technology and culture; and
- An understanding of foreign languages and regional studies.

Strength in these four areas cannot be achieved only within the context of formal educational programs. Institutions offering undergraduate training necessarily focus their attention and limited resources on developing a student's basic skills, understanding, and intellectual outlook needed to maintain professional success over the course of several decades. Real work experience is an indispensable element of education and training for international construction.

PROGRAMS OF STUDY

Education in civil engineering or architecture is the primary course of study for professionals entering construction and design leading to construction. Other engineering and scientific disciplines, culture, history, art, and the often intuitive processes of design are also essential elements of knowledge for the construction professional. However, construction professionals note that construction cannot be taught in the same way as manufacturing or other activities with standardized production. Construction training needs the specificity of carefully chosen cases to balance the tendency of formal educational programs toward abstraction and generalization. Despite Japanese and European experience with modular housing, the failure of Project Breakthrough in the early 1970s was an example of the mistaken belief that structures could be built the same way as machines (i.e., using the mass production lines of the automobile industry).

Engineering

In the United States today, 267 academic institutions offer 1,323 engineering programs accredited at the bachelor's level and 30 at the master's level (see Table 9). (While there are a much larger number of graduate programs they do not require accreditation.) The general criteria for basic accreditation of engineering programs require at least one year's training in a combination of mathematics and basic sciences, one year's training in engineering science, one-half year's training in engineering design, and one-half year's training in humanities and social sciences. Up to one year is then available for other required and elective courses. The criteria for accreditation at the advanced level require the completion of a basic level program, plus a fifth year. In the additional year, at least two-thirds must comprise some combination of advanced level work in mathematics, basic science, engineering science, and engineering design. Table 10 provides a perspective on the annual number of graduates of engineering programs at the B.S., M.S., and Ph.D. degree levels.

There are four accredited programs in engineering management. However, engineering management programs are typically offered at the M.S. degree level, and accreditation at the graduate level is not prevalent because of a restrictive policy which severely limits the accreditation opportunities for engineering programs at that level.

TABLE 9 Total Accredited Engineering Programs by Program Area, as of October 1986

Program Area	Bachelor's Level	Master's Level[a]
Civil, construction	201	1
Engineering management	3	1
Architectural engineering	10	0
Mechanical engineering	218	2
Electrical engineering	238	3
Chemical engineering	142	1
Industrial engineering	85	1
All other (24 areas)	426	21
Total	1,323	30

[a]These are the accredited programs at the master's level. Most accreditation occurs at the bachelor's level, so that there are only a few graduate programs counted for accreditation purposes.

TABLE 10 Degrees in Engineering Awarded, 1986

Program Area	Degree		
	B.S.	M.S.	Ph.D.
Civil engineering	8,798	3,197	439
Engineering management	(N/A)	(N/A)	(N/A)
Architectural engineering	381	48	0
Mechanical engineering	16,702	3,462	565
Electrical engineering	24,514	5,926	779
Chemical engineering	6,148	1,430	534
Industrial engineering	4,645	1,798	120
All other (15 areas)	16,990	7,164	1,249
Total	78,178	23,025	3,686

Note: Currently, there are 95 institutions offering four-year bachelor's level programs, and 155 offering two-year associate degree programs in engineering technology. These institutions offer 273 and 460 programs at the four-year and two-year levels, respectively. No published data are available on the number of degrees currently awarded per year in the technology programs.

At present, there are approximately 20 graduate level programs in engineering management offered at U.S. institutions.

Architecture

There are 103 accredited professional architecture degree programs in North America. A professional degree is either a five-year bachelor of architecture or a master of architecture. According to statistics that are available from the National Architectural Accrediting Board (NAAB), 3,088 B.Arch. degrees and 1,545 M.Arch. degrees were conferred in 1986–1987.

Although it has been said that "there exist as many curricula as there are programs in architecture, and in many schools there are a number of options that lead to the completion of the degree requirements," professional architecture programs actually share similar core curricula. Criteria for accreditation require courses in design, history, materials, human behavior, practice, and so on, with the emphasis placed on the design studio. Other courses taken either within the architecture school or in other departments are meant to complement and enhance the design core of the program. The philosophy statement of the Graduate Program in Architecture at Columbia University is representative of many architecture schools in its declaration that

> Columbia's Graduate School of Architecture is dedicated to the proposition that architectural design has always been and will continue to be the core of professional education. Behavioral, technological, and art historical course work is offered as support for the design studio. Often attempts are made to integrate the attitudes inherent in these disciplines into the design exercises. However, it is the ability to synthesize vast, differentiated bodies of knowledge as they affect and modify the design decision-making process that is stressed.

In addition to the basic core requirements, there may also be a sequence of courses in architectural history and theory. Many schools also require, or at least encourage, speech or writing courses, and other work in the humanities and social sciences to parallel professional courses.

Engineering and Architecture Technology

Engineering and architecture both involve a mix of technical skill and creative application of judgment about how general principles apply in specific cases. The relative balance between innovative thinking and straightforward analysis shifts from job to job, and from task to task within a project. Opportunities arise for dividing the labor, giving rise in turn to opportunities for personnel who function as a bridge between designer and craftsman.

Engineering and architectural technology require the application of scientific and engineering knowledge and methods combined with technical skills in support of engineering and architecture. The technologist is applications-oriented, building on a background of applied mathematics, science, and technology to produce practical, workable results quickly; to install and operate technical systems; to devise hardware from proven concepts; to develop and produce products; to service machines and systems; to manage construction and production processes; and to provide sales support for technical products and systems.

Normally, the technologist will hold a degree from an accredited engineering or architecture technology program. In contrast to the two-year programs of training for technicians qualified to conduct relatively standard field measurements and laboratory tests, the technologist may spend an additional one to two years receiving training in basic principles. Because of his key role as an implementer, the technologist is called on to make independent judgments that will expedite the work without jeopardizing its effectiveness, safety, or cost. The technologist should be able to understand the components of systems and be able to operate the systems to achieve conceptual goals established by the responsible engineering or architecture professional.

Continuing Education

Generally speaking, professional enhancement through continuing education probably offers the most promise for the near-future development of professionals in the international construction field. Many opportunities exist for both architects and engineers, with diverse subject matters, institutions, and lengths of course. The subject matter may range from technical topics at a high level of sophistication, to administration and management. Continuing professional education courses are offered primarily by educational institutions, professional and technical societies, large corporations, and engineering firms.

The policy of the American Institute of Architects (AIA) on continuing professional education states:

> The ultimate responsibility for professional development lies with the individual architect. Professional development occurs properly in both formal continuing education and less formal learning experiences, including everyday professional practice. The AIA advocates the professional development of its members and is committed to provide resources and services in its support.

In many instances, course offerings in subject areas frequently not included in formal, university-based education are more properly available in a continuing professional education setting. Economics, cost estimating, real estate principles, management, and other courses that are not traditionally offered in a professional degree program might in fact have greater impact on the professional student who already has some work experience on which to build.

ISSUES IN CIVIL ENGINEERING

Civil engineering teaching in the past two decades has focused on methods of analysis. The emphasis has been on fundamental studies of mechanics, applied mathematics, and the analysis of structures or of systems. The computer has already influenced much of this teaching and that influence is increasing. Courses in steel and concrete structures do include current practice as expressed in codes and do focus on the principles of detailed proportioning once the form and loads are given, but civil engineering education is almost exclusively analytic, concentrating on instilling basic knowledge and depending on subsequent on-the-job experience to teach students how to apply this knowledge.

Emphasis on Design

This dominance of analysis means that there is almost no teaching devoted to design as a synthesis, to construction as the process of economical building, and to the performance and permanence of civil works as derived from field observations. The present thrust of education treats the works as objects for analysis rather than as subjects for creating new ideas in design and construction.

This teaching direction mirrors closely the present state of structural research weighted heavily toward methods of analysis and on computers. From a design standpoint this research is passive; it is oriented toward improvements in analysis rather than changes in design. It is certainly true that more efficient analysis can help design as a process, but it is hard to show how more competitive construction has arisen because design as a process is more efficient.

Design improvement depends on effective performance evaluation, for which field observations are crucial. The performance of actual civil works is not currently a part of education. Existing courses on analysis do provide a sound basis for interpreting the results of field observations, but at present this powerful potential is unrealized. Because educators do not use data from real works in their teaching, there is little effort to collect data, and such data, if collected, rarely are published. In the teaching of concrete structures, for example, performance criteria are taught almost exclusively through code provisions, which is of course necessary but not sufficient. Design is treated as controlled by a generalized set of rules rather than as informed by specific, but characteristic, examples.

One dramatic example of the tendency to treat improvements in civil works as arising from general analysis is the highway pavement research program pursued by the Bureau of Public Roads from 1920 to 1945. The program tried to represent observations of performance and constructed a long series of laboratory analytic studies devoted to the fundamentals of pavement design. That work ultimately had to be abandoned, and following World War II the bureau returned to a major full-scale field study as the basis for design improvements. However, the lessons learned in this analytical work still underlie current understanding of the materials mechanics of pavement behavior.

Construction Management

Construction and construction management are treated separately from design, and the ultimate relationship between design and construction is only rarely discussed. Also, construction is often treated as a set of processes to be analyzed and not as a series of individual unique cases. This distinction is well recognized in practice and was recently articulated in an editorial of the *Engineering News Record* (May 7, 1987) commenting on an April 1987 workshop sponsored by the National Science Foundation at Lehigh University:

> The dominant theme was rejection of task-specific robots or expert systems and embracing new methods to distribute information. The reason for both is the same: construction is one of the messiest industries around. Each project differs from others and each changes from day to day. Figuring out how to complete a construction project efficiently has no relation to figuring out how to make the same

spot weld in the same car chassis every few seconds. Repetition is irrelevant; coordination is vital.

The absence of construction management in a civil engineering education thus leaves a distinct void that has important implications for how effectively the resulting professionals can support construction. This void may then influence the entire U.S. construction industry.

ISSUES IN ARCHITECTURE

For most of recorded history (going back to the time of the Egyptian pharaohs) architects were educated by becoming apprenticed to a professional already in practice. Toward the end of the nineteenth century, formal schools were established outside of the architect's office to provide special "ateliers" for gaining a professional education, usually with more emphasis on the art of architecture than was afforded an apprentice in a normal office. The Ecole des Beaux Arts in Paris became by the early part of the twentieth century the leading place to study architecture, if one's family could afford it. The influence of this school's method of teaching spread around the world as colleges and universities began to offer architecture courses within their programs.

By the time of World War II, the "design studio"—a group of 10 to 20 students under the dominance of a "crit" (member of the teaching faculty who criticized the students work)—formed the heart of all schools of architecture, and still does to this day. The method of teaching is essentially, therefore, still a form of apprenticeship, but with a series of masters and with no visible financial connection between master and apprentice. When a student is fortunate enough to work with one or more really skilled crit, the educational format is superb, but when the crit is neither a skilled practitioner nor a good teacher (and this is all too often the case in the past few years), the student is not well educated.

Further, the context of the university tends to be dominated and overshadowed by the demands of the design studio. It is a common sight to find the lights in the architecture design studios burning all night when a student project is coming up for juried evaluation. The term "en charrette" was coined during these periods of intense concentration in the Ecole des Beaux Arts, because at the final and formal end of a project's schedule, a cart or "charrette" would be pulled through the studios to collect the students' work.

Critics of the architecture education system claim that the design studio is overemphasized, and that technical instruction suffers as a result. Proponents of the status quo suggest that it is possible for a student to acquire needed specific technical skills in the workplace.

Specialization and Small Practices

While education emphasizes creative design in the studio setting, architecture as practiced by professionals who are licensed by each state is a synthesizing activity which converts the requirements of a client into building spaces that are structurally sound, provide a safe and healthy environment, are economically suited to the client's needs, and are stylistically in keeping with both the client's tastes and the professional community's standards. In practice, the architect will use consulting engineers for such specialized design and analysis as structural systems, heating, ventilating, and air-conditioning systems, lighting, acoustics, energy efficiency, cost estimating, and so on. The practicing professional then must be prepared during his educational program to understand and communicate with such consultants, but not necessarily to have these analytic skills himself.

As the novice architect moves into practice, he will be led to focus on design, engineering, construction, or the production of working drawings and specifications. In large firms most employees will end up on the production side, and it is those skills that are addressed in varying degrees by schools of architecture. Those schools that are organized around two- or four-year programs of "architectural technology" are most clearly focused on providing the training for people who will devote their careers to design production. There are about a dozen schools in the United States with programs in architectural engineering, having as their purpose the preparation of professionals who will focus on that aspect of practice (which almost always means structural engineering, however). The vast majority of students are being educated as though they will be designers. It is not surprising that in the United States so many small architecture firms exist (the median size of architecture firms is 4.2 persons on the staff), since the designer has to be seen as "gifted" to play that role in a large firm, and not many graduates of architecture schools can meet the criteria associated with "gifted."

In the past few years, especially in high-priced real estate areas such as California and New York, architecture graduates have been pursuing careers in the development side of the building industry.

This is not only a more lucrative career choice, but since a development firm normally builds first and sells or leases later, it provides an easier means to being "the designer" for the projects. As indicated elsewhere, international practice (outside of the United States) tends to be dominated by firms that are vertically integrated so that architects and engineers are staff members of large firms that provide "turn-key" services to their clients, a situation poorly suited to the great majority of U.S. architectural firms.

Architectural Research and Education

The 103 schools of architecture in North America have had an inconsistent history of research. The majority of the schools have no formal unit concerned with research, although individual faculty members might undertake research studies themselves. It would be unusual for such individual research efforts to include undergraduate students, and it would be difficult to document the contributions that such research makes to the teaching program of the school. The Architectural Research Centers Consortium, created in 1976, is a group of some 30 research units attached to schools of architecture that provides a means of exchanging research plans and results. The consortium, originally intended to make it possible to undertake large-scale research efforts by combining institutions into teams, has had limited success.

To enhance prospects for architectural research as a contribution to education, the committee recommends the inclusion of advanced technological content in the architectural curriculum:

- Courses that provide an understanding of how buildings are actually built, not just the materials and equipment that go into a building, but the tools and techniques used for construction in the field and in the factory;
- Courses that provide working experience in the use of computers as tools of design and analysis; and
- Design studio courses organized around making use of the growing research base, which ranges from research on human needs to research on indoor air quality.

Unless professionals in practice receive early training in how to use the knowledge base available to them, they will not likely do so. Continuing education programs should be offered by the AIA and professional schools to provide professionals in practice opportunities to learn about, and experience, the design and construction practices of other countries.

SKILLS FOR GLOBAL ENTERPRISE

Knowledge of foreign languages, cultural environments of other nations, and significance of historic and cultural characteristics for both business practice and building is the basis for effective performance within the international construction community. The committee observes that the United States has not kept up with its foreign competition in developing these skills.

For example, a few architectural schools offer study-abroad programs, but faculty opinions about the value of these programs is mixed. In most cases, students returned to their home institutions with a more sophisticated and heightened awareness of design possibilities; in one or two instances, this held true for building technology problems as well. Most of the programs provide for ample opportunity for mixing with students, faculty, and local practitioners in other countries.

While it is generally agreed that this type of experience contributes significantly to any education, most faculty responding to a questionnaire distributed by the committee felt that such a program did not adequately prepare students for professional involvement in international construction projects. There was uniform agreement that few students had foreign language skills, with the exception of a program in China organized by Carnegie-Mellon University, for which the students were required to study Chinese for one year prior to enrolling in the program.

Cross-Cultural Training for the Construction Industry

The "Agenda for American Competitiveness," issued by the Business-Higher Education Forum, the Northeast-Midwest Coalition, and the Congressional Clearinghouse for the Future, points out that about 10,000 English-speaking Japanese business executives work in America, handling billions of dollars in trade, some of that trade in construction. However, very few of the 1,000 American businessmen in Japan can speak Japanese. Generally speaking, the Japanese seem to be much better equipped to come to the United States to study the technologies and practices of industry here than Americans would be to go to Japan.

A 1984 survey cited by the "Agenda" report demonstrated that U.S. students are increasingly ignorant of world geography (and in particular, countries which are of strategic importance to the United States): fewer than half could locate Iran, only 30 percent could locate Afghanistan, and only 25 percent knew where El Salvador was.

The "Agenda" report recommended that colleges and universities significantly strengthen their international studies courses—language, cultural, political, and economics—and make them readily available to U.S. business executives as part of their own lifelong learning programs. Certainly business degree programs cannot afford to ignore the increased globalization of business both domestically and in overseas markets.

Engineering schools would do well to consider foreign language degree requirements and international studies courses. One way to involve engineering students in such studies would be to design the courses so that they would include engineering aspects of other cultures and an emphasis on the relationship between technology and culture. Architecture students already receive a strong dose of cultural studies in the architectural history courses that are a part of the core curriculum. The committee suggests in particular that the Fulbright Program should be expanded to encourage more architects and engineers to gain exposure to other cultures.

According to a report* of the National Academy of Engineering (NAE), one way of connecting U.S. engineers with foreign technologies is by increasing their participation in international activities, particularly in the setting of international standards for products and services. Another way would be through the development of case studies, researched and written by expert consultants from various nations, and then incorporated into the curriculum.

The NAE report stated that "technological isolation will surely undermine the future of our industries." Increasingly, as the U.S. design and construction industries look to greater participation in the global enterprise, engineering schools, professional societies, and business organizations must look outside themselves to learn how to do business in an international economy. Only through more deliberate exposure to foreign languages, geography, business, and culture will U.S. design professionals gain access to foreign-originated

---

* *Strengthening U.S. Engineering Through International Cooperation: Some Recommendations for Action*, National Academy Press, Washington, D.C., 1987.

technologies, foster cooperation with foreign work forces overseas, and develop an increased ability to deal with foreign sources of business opportunities and finance.

Acquiring Foreign Languages

The study of foreign languages is not generally of concern in the present educational programs of either architects or engineers. In the context of this study, well-developed language skills are far more important for the comprehension of a particular culture than for the purpose of doing business, as English has virtually taken over as the international language. According to a Modern Language Association (MLA) Language Enrollment Survey conducted in the fall of 1986, total enrollment in languages other than English at American colleges and universities exceeds 1 million for the first time in 14 years. The survey results indicate an increase of almost 4 percent between 1983 and 1986, continuing a trend that began in 1980. It is interesting to note that Japanese and Chinese courses had the fastest-growing enrollments (45 and 28 percent increases, respectively), although total numbers of students still trail those studying Spanish, French, German, or Russian. Figures are not, however, broken down according to fields of study.

In the NAE report, the committee stated, "Educational institutions should respond to the urgent need for increased capability in Asian languages and culture for U.S. engineers and technologists. Graduate degree programs in engineering and applied sciences should emphasize the need for spoken and technical competency in at least one foreign language." The same chapter also emphasizes "the usefulness of early study of languages and experience that reinforces language skills needs to be better appreciated by young people who wish to pursue careers in engineering and technology." It is also suggested that study of any language be done in conjunction with study of the technology and the culture in question. Having some familiarity with a foreign culture, even without the language, can be very helpful to those professionals working overseas.

As far as architecture programs are concerned, it is safe to say that the same is true, although this information is school-specific, according to the Association of Collegiate Schools of Architecture (ACSA). For certain special areas or programs within the field, such as architectural history or theory, certain languages may be mandatory (such as a Rensselaer Polytechnic Institute program in Rome, which has a prerequisite of one year of Italian), but languages are not required for general admission to most schools. In undergraduate architecture programs, there is generally more time for elective courses than in engineering schools, as the first professional degree is often structured for five or six years of study. In many architecture programs, students are strongly encouraged but not required to take courses in other languages.

Several factors favor the study of foreign languages by architecture and engineering students:

- Language skills are transferable; once one foreign language has been studied and/or mastered, it becomes much easier to tackle another one, because those particular mental skills have been developed and exercised;
- In the study of a language, the student learns something about the culture of that nation, which can be very useful professionally; and
- Language skills enhance a student's accomplishments, making him more marketable to international projects in the building industry.

Factors that work against the study of foreign languages have largely to do with time limitations. Since both engineering and architecture programs are fairly highly structured, and since in many cases the languages that have been studied will not be useful in currently developing international markets, continuing education courses may be the answer for the short term.

INTERNATIONAL PROJECT MANAGEMENT

Most training and education in international project management have been through hands-on experience obtained individually by members of construction and engineering firms engaged in executing individual projects. The international market is characterized by a number of unique conditions that can dramatically affect project cost, schedule, and quality. These conditions are very country- and site-specific, and substantial local market research is required of prospective engineering and construction firms seeking overseas work.

American firms performing overseas construction work may have difficulty in obtaining the required commercial licensing, face possible transportation delays, and encounter difficulties in obtaining customs clearances. These issues may be further compounded by language barriers and in-country restrictions on the employment of Americans.

Firms entering the international marketplace need to be ready to react to unique labor laws, requirements for use of local materials, and significant differences in quality standards. They must also be prepared to make substantial investments in technology transfer and training to develop the skill base necessary to complete the project. The host country's business practices and construction process concepts are frequently at odds with a U.S. contractor's normal business methodologies and procedures.

Personnel conducting contract negotiations frequently do not have sufficient time to become totally familiar with nuances of the construction process of a given country. Expensive lessons have been learned simply because inexperienced contractors have failed to take into account the impact of the host country's culture on their standard operating procedures.

The degree to which the client becomes involved in the construction project can be a positive or negative factor depending on his familiarity with the construction process and the role he chooses to play. If the client chooses to act as liaison with the host nation's government, the contractor will be at the mercy of the client when it comes to acquiring needed information, permits, and other approvals. This can have a significant impact on project schedules and costs.

In general, the committee concludes that U.S. construction and design firms and their professionals need better training for their role in the global economy. The relationship built between Philipp Holzmann and J. A. Jones Construction Company (see Case Study 5) illustrates how, with capable people willing to cooperate, firms from different nations can work together to their mutual benefit. The motivation for this inquiry has been competition, but it is becoming clear that cooperation is equally important. On both counts, U.S. skills are lagging.

CASE STUDY 5: BUILDING INTERNATIONAL RELATIONSHIPS: PHILIPP HOLZMANN AG AND J. A. JONES CONSTRUCTION COMPANY

A large German firm's acquisition of a large American construction firm nine years ago is called a successful marriage by the participants. Not only have the two management styles mixed well, they say, but the transfer of technology between Philipp Holzmann and J. A. Jones Construction Company has benefited both companies by allowing expansion of worldwide construction horizons.

Philipp Holzmann AG had grown from a small family business into a leader in international construction. Founded as a railway contractor near Frankfurt am Main in 1849, the firm quickly extended activities to include civil engineering and building of all types. Holzmann won its first major contract outside Germany, the main railway station in Amsterdam, in 1882, and since then has been active in many European countries, South America, Asia, and Africa. Holzmann now has major activities ongoing in the United States.

The company designs and builds public and commercial buildings, manufacturing and industrial plants, marine structures, and mass transit facilities. Its range of services includes turn-key projects as well as maintenance and operation of facilities. Holzmann also undertakes reconstruction and modernization of buildings and industrial plants.

The general management and overseas departments of Philipp Holzmann are headquartered in Frankfurt, West Germany. The company operates 30 branch offices throughout Germany and has more than 50 domestic and foreign subsidiaries engaged in special fields of construction and construction-related activities around the world.

Holzmann is represented in the United States through its subsidiary, Philipp Holzmann USA, by Jones Group, Inc., in Charlotte, North Carolina, and by Lockwood Greene Engineers, Inc., Spartanburg, South Carolina, in addition to other subsidiaries. Jones is a construction contractor, and Lockwood Greene represents the architectural and engineering side.

In 1986 the Holzmann group of companies had sales of $6.6 billion worldwide. Approximately 48 percent of that total was in the United States. The decision to enter the U.S. construction and engineering market was a result of events around the world. Since its first international experience in the late nineteenth century, Philipp

Holzmann AG has set corporate strategies beyond German boundaries. In the earliest years, the company built the legendary Baghdad railroad, the Dar-es-Salaam railroad in Africa.

By World War I, Holzmann had built the first skyscraper in Buenos Aires and entered the U.S. market with construction on the Barge Canal in New York. More recently, its activities centered in the Middle East—where some of the world's largest construction projects have been built with oil revenues—including hospitals and a sports stadium in Saudi Arabia. In the early 1970s, more than 50 percent of the company's foreign revenues came from Saudi Arabia. But the Iranian revolution, Iran-Iraq war, and the softening of oil prices made the prospects of a blooming construction market in the Middle East seem less promising.

Holzmann carefully evaluated the possibility of future construction market collapses and, in order to protect the company from such uncertainties, decided to diversify by investing in other countries. Economic and political stability and a self-sustaining market in the United States were an attraction. Holzmann sought a U.S. company that would complement its strengths and, in 1979, acquired the Charlotte-based J. A. Jones Construction Company which had 90 years of experience in the U.S. construction industry.

J. A. Jones has its own history and many successes. Founded in Charlotte, North Carolina, it has grown to become a U.S. and international leader. James Addison Jones started his work as a bricklayer in 1890, and got most of his early experience building for the textile industry throughout the southern states. But J. A. Jones's first project, like Holzmann's, involved the railroads. Jones built the dining car facility in Charlotte for the Southern Railroad Company.

Following the 1930s depression, Jones signed one of the largest construction contracts to that date, for a new airbase in the Panama Canal Zone. Since that project, military construction has been an important factor, including construction of Liberty ships during World War II, followed by what was then the "largest construction project in the history of the world"—the gaseous diffusion plant at Oak Ridge, Tennessee.

Following World War II, Jones began a long series of heavy and highway construction work while continuing commercial building throughout the country. Today, the company is also involved in industrial and energy work as well. As it, too, looked to the Middle East for work in the 1970s, J. A. Jones competed against Holzmann;

they then worked as a joint venture on a military training center project in Saudi Arabia.

In October 1978, J. A. Jones Construction Company announced its agreement to be acquired by Philipp Holzmann AG. The purchase ended a plan by Jones to be employee-owned, a process begun in 1968. Stockholders were assured that the firm's name, management, and work force would not change. Today, Holzmann is represented only on the board of directors.

At the time of the purchase, Jones's stock price was valued at $23.06 per share, which was determined by the company's own estimate of its value, since the stock was not widely traded. Holzmann's offer amounted to $40.61 per share. Although it was stated that several other companies had an interest in the purchase of J. A. Jones, Holzmann's offer was accepted. According to Johnnie H. Jones, then executive vice-president and now chairman and president of Jones Group, Inc., "The philosophy and integrity of Holzmann's management were most compatible with the Jones team."

Three primary reasons were given for the marriage: (1) the financial strength of Holzmann would enable Jones to resume its growth and continue to grow faster; (2) the combined international experience of the two companies would improve their competitiveness in foreign markets; and (3) the merger would allow Jones access to the larger firm's technology, with the combined international experience of the two companies improving their competitiveness in foreign markets. At that time Holzmann spent more than $3 million a year on research and held several patents in concrete technology.

Over the nine years since acquisition, the Holzmann-Jones partnership has allowed both companies to bid on a greater variety of projects because of broader market presence and shared technologies. In addition, the financial strength of Holzmann has given Jones bonding capacity to increase its volume of work and the size of its projects.

Benefits of the merger surfaced early. Jones became more competitive in heavy construction, where Holzmann had for decades been a world leader, and entered the marine field with a sunken tube tunnel contract and one for a floating pontoon bridge. Holzmann gained expertise in the chemical plant market and in high-rise construction, long a J. A. Jones strength but at the time a costly type of construction in West Germany.

To diversify further in America on the design side of construction, Holzmann in 1981 purchased 80 percent of Lockwood Greene

Engineers, Inc., of Spartanburg, South Carolina. Holzmann also encouraged Jones Group's formation of a new service company, which specializes in facilities management, similar to one Holzmann founded in the early 1980s.

The two companies have set up an informal employees program that enables young engineers to travel and work somewhat like exchange students abroad. Management level staff members also take part in orientation programs between West Germany and the United States.

J. A. Jones has added a dimension in international construction that has benefited the parent company. By offering procurement services for materials for projects where many of the designs call for American standards, Jones can help Holzmann avoid problems of selecting goods in a foreign country.

In turn, Johnnie Jones says that Holzmann "does not interfere with our operation but provides support. Keeping our management intact proved to us that Holzmann agreed with our philosophy that people are our most important assets."

The Jones Company now can take on major heavy construction projects which heretofore would have been undertaken only in a joint venture. And in a totally new direction, Jones signed its first contract to build, own, and operate a lignite mine in Louisiana. "It required a substantial investment in the beginning, but we wouldn't have been able to do it without the financial support of Philipp Holzmann," says Jones.

Both companies are in the process of diversifying in technical fields, expanding in other locations, and reestablishing positions in the international market. The formation of a real estate development company in Atlanta, Mark III, and additional activities from Queens Properties, Inc., in Charlotte, were steps in diversification.

Because the financial capabilities of construction firms have new importance, Jones Group this year formed Jones Capital Corporation to develop project financing and to hold the assets of projects in which Jones is maintaining an ownership position.

While still run as separate entities, Holzmann and Jones combined can pursue the largest construction projects in the world.

6
Pursuit of Innovation

The committee observes that the United States and the world are experiencing rapid technological advance, but that applications to construction have been relatively limited. U.S. construction and design have in the past played an important world leadership role that is now threatened, in part due to society's growing willingness to assign liability on a basis of ability to pay, and in part due to the competition's commitment to progress. In addition to making a greater effort in research and development and enhancing education and training, the U.S. construction industry must rekindle its enthusiasm for innovation if it is to maintain its place in the global economy.

NATURE OF INNOVATION

Innovation can occur in a ***design*** (e.g., suspension bridge) or in a ***material*** (e.g., reinforced concrete). It can occur by a ***major breakthrough*** based on a novel invention that dramatically and suddenly changes what we build (e.g., the need for airports was created by the invention of the airplane). However, innovation is more frequently achieved through many ***incremental improvements*** that serve to make a technology useful (e.g., improved roadway paving materials).

Most innovations (including those in infrastructure) are the result of absorbing an invention, often after it has been developed for another purpose. In each era, certain primary inventions become the basis for much of the innovation that occurs. At present, a number of new primary inventions are driving change in construction:

- Photonics: those inventions that produce coherent light that can be amplified and propagated, such as lasers, masers, and fiber optics. Paths of light will increasingly replace the wires along which messages flow, and lasers have found application in field surveying as well as in factory cutting and welding.
- Biotechnology: genetic engineering, neuroengineering, and the recoding of macromolecules of living things to produce new organic substances that can have applications in buildings and infrastructure. Pollution control and hazardous waste disposal stand to benefit greatly.
- Materials science: fundamental reformation and fabrication of inorganic materials to provide performance characteristics not found in nature, such as high-strength composites, rapid-flow membrane technology, and superconductivity. This latest discovery may have far-reaching impacts on the storage of electricity and transport technology.
- Microelectronics: circuits, switching mechanisms, data storage devices, amplifiers, and sensors. Such devices can extend human strength and dexterity through robotics; support data collection and analysis to enhance the speed and effectiveness of human actions; and make possible graphic input and output of data and so begin to substitute a picture for a thousand words.

Such innovations may have profound implications for construction. They may change working relationships between designers and constructors. Buildings themselves are becoming more "intelligent" as they have electronic enhancements added to their information and communications systems as well as the controls for mechanical equipment. Robotics and other forms of automation are beginning to provide practical applications for performing difficult or dangerous job-site tasks, and may well alter the economics of much work done on the job.

The impressive efforts of Japan's construction industry have been described. European firms as well have undertaken aggressive searches for innovation, particularly in the development of proprietary construction systems. These new systems are based on extensive integration of design, fabrication, and erection processes, all of which are carried out by a single firm. Several large European firms have succeeded in vertically integrating their business structures to include the management of key materials supplies, design and engineering expertise, development know-how, and financing capability. The economic integration into one firm of these functions more easily allows the construction firm to capture the economic benefits of productivity and quality improvements through the adoption of new technologies. Because there is better control of costs, such integrated firms develop a competitive advantage.

Firms in the United States continue to take a passive attitude toward construction innovation. Even the largest U.S. firms, which may have the resources to undertake significant research programs, continue to put their faith in the strategy of being "technology followers." Indeed, many large U.S. construction firms have suggested that by not being committed to any one proprietary technology they are at a strategic advantage in being able to pick and choose among the latest technologies around the world. The committee questions the wisdom of this strategy. In a global market, those firms that have developed a proprietary technical advantage are in a position to refuse to grant licenses to firms with which they do not wish to compete. Even when the technology is available in principle, individuals and firms are often deterred by the initial intellectual and financial investment required to apply it in practice.

There are three ground rules that seem to be needed for any serious effort to encourage innovation in the U.S. construction industry:

- For major innovations to take hold and become common, they need to be founded on a confluence of basic research and practical improvements. In other words, they rely as much on basic research (to the extent it is still useful to use that term) as they do on applied engineering. Often the area of basic research used bears no obvious relation to the eventual practical application.
- The search for innovation must allow for major breakthroughs followed by incremental advances, and it can include improvements in design as well as materials. During the overall process of invention, various improvements reinforce one another and encourage public demand, in such a way as to promote further innovations.
- Mechanisms are needed to capture the potentially sizable payoffs of innovation for those who attempt it. Early American bridge innovations (from the nineteenth century) are a good example, because generous royalties were paid for the use of the ideas that had been granted patents.

OPPORTUNITY IN INFRASTRUCTURE

Within the United States, as in most of the world, there is without question an opportunity to increase the performance characteristics of those systems used to transport people and goods, obtain water, remove wastes, supply energy, and facilitate communications. There is also reason to include those buildings used either for public purposes (e.g., schools and hospitals) or built with public funds (e.g., government offices, court houses, and prisons) as a part of the public works infrastructure.

Under this broad definition of infrastructure, the United States in 1984 invested $102 billion, 30 percent of its design and construction budgets (see Tables 11 to 13).

Development of advanced infrastructure is a challenge, worthy of a cooperative international effort. It will be difficult to structure these developments to match the performance requirements of a society utilizing advanced science and technology, and make them more than incremental improvements to the present modal technologies. In the developing part of the world, where the most rapid urbanization is happening, the challenge is to develop technology appropriate to their requirements rather than to impose solutions produced for industrial nations.

There are two reasons for the United States to do more about advancing the technology of infrastructure. It would benefit within its own borders from new and higher-performance systems, and it could also have another opportunity for marketing its technology on a global basis. This committee recognizes the urgency of maintaining and extending the existing networks of public works that underlie U.S. cities. However, the nation also needs to develop new and higher-performing technologies to gain the potential market that improved performance makes possible and to avoid an indefinite future drain on the public purse from maintaining the older systems.

The existing infrastructure is based on a set of inventions that emerged toward the end of the last century. These inventions produced a second generation of urban systems that provided performance characteristics substantially different from those previously used in all of human history:

- Structural steel frames for buildings. When this method of construction was first introduced in the 1880s in Chicago, it made it possible to erect structures that were taller than the five- or six-floor

TABLE 11 Estimates of Private Construction Volume that Might Be Included Within the Category of Infrastructure (in $million), 1984

Type of Private Construction (by census category)	Total Value[a]	Infrastructure Value[b]
Residential buildings	145,059	
Nonresidential buildings (organized by functions)		
Industrial	13,745	
Office	25,940	
Other commercial (warehouses, silos, retail stores shopping malls, drugstores, parking garages, service stations, barber shops, dance schools)	22,167	10,000
Religious	2,132	
Educational	1,411	1,411
Hospital and institutional	6,297	6,297
Miscellaneous (movie theatres, casinos, health clubs, radio and television stations, including bus and airline terminals, public utility buildings)	2,455	490
Subtotal	74,147	18,198
Farm nonresidential buildings	2,860	
Public utilities (organized by industries)		
Telephone and telegraph	7,174	7,174
Railroads	3,671	3,671
Electric light and power	19,473	19,473
Gas	3,233	3,233
Petroleum pipelines	271	271
Subtotal	33,822	33,822
All other (privately owned streets, bridges, parking areas, dams, reservoirs, sewer, water facilities, parks, and playgrounds)	1,912	1,912
Total[c]	257,801	53,932

Source: Bureau of the Census data, with staff extensions, 1984.

[a]Value includes cost of materials, labor, equipment rental, contractor profit, owners' overhead costs, architect and engineer services, miscellaneous charges on owners' books, interest, and taxes during construction.
[b]Infrastructure is defined as including all buildings used for public purposes (e.g., schools) whether paid for privately or publicly, and all construction of "networks" for supporting buildings (e.g., roads). Where exact data are not provided an estimate has been made.
[c]Subtotals may not add to totals because of rounding.

TABLE 12 Estimates of Public Construction Volume that Might Be Included Within the Category of Infrastructure (in $million), 1984

Type of Public Construction (by census category)	Total Value[a]	Infrastructure Value[b]
Buildings		
(by functions)		
Housing and redevelopment	1,636	
Industrial	1,828	
Educational	5,557	5,557
Hospital	2,039	82,039
Other	6,822	6,822
(administrative; police, fire, bus, and streetcar stations; subway garages and barns; jails; parking facilities; airport and marine terminals; electric power generating buildings; and so on)		
Subtotal	17,883	14,418
Highways and streets	16,294	16,294
Military facilities	2,839	
Conservation and development (water resource protection and control, fish hatcheries, spillways, pollution control, levees, seawalls, canals, docks, piers, wharves, berths, and reservoirs built other than for potable water supply)	4,654	4,654
Sewer systems	6,241	6,241
Water supply facilities	2,621	2,621
Miscellaneous (recreational facilities, power-generating facilities, and other open construction for subways, streetcars, airport runways, parking, and so on)	4,654	4,654
Total[c]	55,186	48,882

Source: Bureau of the Census data, with staff extensions, 1984.

[a]Value includes cost of materials, labor, equipment rental, contractor profit, owners' overhead costs, architect and engineer services, miscellaneous charges on owners' books, interest, and taxes during construction.
[b]Infrastructure is defined as including all buildings used for public purposes (e.g., schools) whether paid for privately or publicly, and all construction of "networks" for supporting buildings (e.g., roads). Where exact data are not provided an estimate has been made.
[c]Subtotals may not add to totals because of rounding.

TABLE 13 Estimates of Private and Public Construction Volume that Might Be Included Within the Category of Infrastructure (in $million), 1984

Type of Construction	Total Value[a]	Infrastructure Value[b]
Public and private[c]	312,987	102,184
Private sector buildings		18,198
Public sector buildings		14,418
Total building components of infrastructure[c]		32,616
Privately financed utility systems		33,822
Publicly financed utility systems		34,464
Total utility components of infrastructure[c]		70,200

Source: Bureau of the Census data, with staff extensions, 1984.

[a]Value includes cost of materials, labor, equipment rental, contractor profit, owners' overhead costs, architect and engineer services, miscellaneous charges on owners' books, interest, and taxes during construction.
[b]Infrastructure is defined as including all buildings used for public purposes (e.g., schools) whether paid for privately or publicly, and all construction of "networks" for supporting buildings (e.g., roads). Where exact data are not provided an estimate has been made.
[c]Subtotals may not add to totals because of rounding.

limitation of masonry walls that had dominated architectural design for all of prior human history.

- Elevators for moving people and goods vertically in tall buildings, made possible by the Otis inventions for safety. Elevators replaced stairways that, because they required human energy to ascend, were not practical beyond the five- or six-floor limitation of earlier designs.
- The set of inventions that made possible indoor plumbing devices connected to water and waste systems, which replaced the outhouse, the slit trench, and all of the prior disease-ridden methods of disposing of human waste.
- Central heating systems that, especially when they began to use the fluid fossil fuels of oil and gas, changed the logistics of supplying fuel for heat since fuel no longer had to be manually suppled to separate stoves and fireplaces located throughout a building (and ashes no longer had to be removed from each separate heating device).
- The discovery of electricity, and the subsequent invention of

generators, amplifiers, distribution methods, electric motors, and the light bulb, which substituted for the historical use of candles, whale oil, animal power, and so on.

- The telephone system based on the primary invention of Alexander Graham Bell in 1876 that made voice communications possible across great distances, replacing such ancient methods as town criers, messengers, and mail.
- The automobile, or more appropriately the internal combustion engine, which substituted a device for the conversion of a fossil fuel to useful energy for the animal power used in all of human history.
- The subway, or the underground railway, as first introduced in London, which provided for mass transportation within a crowded urban area, without pollution of the air or interference in the arrangement of buildings.

There are many indications of limitations of the performance capability of this second generation of infrastructure technologies relative to today's demands. Their ability to support the activities of modern industry is sorely taxed. While there is the possibility that the recently completed work of the National Council on Public Works Improvement will stimulate Congress to provide major new support for infrastructure innovation, the committee feels that only through effective public-private partnership can innovation be achieved in practice.

Beyond the obvious plea to be made for increased government funding in the field, the programs of other countries illustrate the value to be gained through true partnership of private and public interests in the U.S. construction industry. This partnership should embrace research and innovation for both domestic productivity and international competitive strength.

For example, projects built with government funds can assume the greater commercial risk involved in adopting innovation, as was demonstrated by the introduction of precast concrete segmental tunnel liners to U.S. transit construction. This technology had been widely used around the world (since 1936 in England), but not in the United States because individual transit companies were reluctant to take the risk of being first. The Urban Mass Transportation Administration sponsored a research and development project to install concrete segments in one stretch of the Baltimore subway, and suddenly this became the standard technology for U.S. transit systems. National Science Foundation projects done in cooperation with the construction programs of other federal agencies could play a similar role for introducing innovations into design and construction.

Precedent also exists for private-public cooperation in competition for international projects. While the Three Gorges Project in the People's Republic of China was not resolved as the team might have hoped (see Case Study 6), the experience is a valuable lesson demonstrating U.S. ability to emulate the institutional arrangements of British, French, Dutch, or Scandinavian firms and their governments.

However, even this precedent is not enough. The U.S. construction industries' 1.2 million firms need a stronger and more effective voice in national policy. Existing industry organizations play an important role in representing the particular interests of their membership, but there is no forum for resolving inevitable conflicts and initiating cooperative activity.

GLOBAL PARTNERSHIP FOR INNOVATION

As the final chapter of this report will discuss, new or altered institutions are needed to make this partnership of private and public interests effective in the United States. The committee feels strongly, however, that the opportunities for innovation in construction and the potential world economic and social benefits of capturing these opportunities warrant partnership on a global scale, a partnership to work in the United States as well as abroad.

U.S. construction and design firms have found it desirable to rely on comparative advantage and pursue a strategy of cooperation rather than competition, as the examples and case studies gathered by the committee have illustrated. The strategy is a good one for innovation as well. To make the strategy work, however, the U.S. construction industry must strive to maintain its traditional leadership in technology, for two key reasons: (1) loss of technological leadership may mean loss of comparative advantage and competitive position and (2) without the strength for good competitive position, meaningful cooperation becomes nearly impossible.

CASE STUDY 6: COOPERATIVE EFFORT BETWEEN U.S. PUBLIC AND PRIVATE SECTORS: PROPOSAL FOR THE THREE GORGES PROJECT IN CHINA

For six weeks in 1985 a group of leaders in engineering design and construction sequestered itself in a hastily assembled office in Washington, D.C. The group's goal was to accomplish a task many might think impossible: create a proposal to design and construct one of the world's largest civil engineering projects—the Three Gorges Project in the People's Republic of China. The impetus for this challenging undertaking was an invitation from high-level Chinese officials for the United States to take a lead role in project development. The enormousness of the Three Gorges Project and the brutal proposal deadline were complicated by the fact that both the proposal and the work would be done through a combination of U.S. private and public sector groups.

The "Team America" effort, as it was dubbed, resulted in much more than a document. The undertaking showed that real or perceived differences and barriers between U.S. government agencies and private firms can be surmounted to meet shared goals. In the case of the Three Gorges Project, where U.S. involvement would have far-reaching effects for the nation and others, the accomplishment was admirable—and one that can serve as a prototype for future cooperative efforts.

Another less favorable, but equally important, lesson came out of this exercise. While the Chinese government accepted the proposal, the work was not pursued due to lack of financial support from U.S. government and/or private industry sources. As a result, proposed feasibility studies are now being done by a nation in which the private and public sectors cooperate to best advantage—Canada.

The Three Gorges Project was conceived early in the 1900s by Dr. Sun Yat-Sen in his "Plan for Industrialization of China." "It is the long-cherished wish for the Chinese . . . to construct the Three Gorges Project. . . . Completion of the project will be of great significance to the industrialization of the country," wrote Sun, who is still hailed as a visionary by his countrymen.

Nearly a century after Sun's predictions, the powerful Yangtze River frequently ravages the valley below with floods that endanger hundreds of thousands of people and major agricultural and industrial bases. Forty percent of China's food supply is grown in this valley. Industry—the heart of the country's revitalization—is crippled as 40 to 60 percent of capacity is idle at any given time due to power shortages. Harnessing the world's third longest river with the Three Gorges Dam would provide approximately 1,300 megawatts of hydroelectric-generating capacity and lead to formation of a nationwide, large-scale electric power pool.

Improving navigation on the river is of significant economic importance to China, and the project would aid passage of ships the size of ocean-going vessels through narrow channels in gorge areas. Three Gorges would be a concrete gravity dam with a crest height of 510 to 575 ft and a length of 7,200 ft. The dam would include two, four-step shiplocks, and the narrow reservoir would back up 100 miles or more.

Following Sun's early vision for the project, plans proceeded slowly over the years due to a variety of changing conditions in China. Pioneering work was done in the 1940s by the Bureau of Reclamation's chief design engineer, Dr. John L. Savage. In the decades of the 1950s and 1960s, the Chinese made a detailed comparison of alternative sites, and in 1979 proposed the currently favored Sandouping site. In 1984, the State Council approved the project's feasibility report and in March 1985, the Chinese completed a preliminary design report.

In May 1985, former Secretary of the Interior William Clark visited China on a diplomatic mission that led high-level Chinese dignitaries to invite the United States to propose a lead role in project development. Clark made a commitment for the United States to aid China by defining concrete steps that could be taken to address technical and financial issues.

Upon his return to the United States in June 1985, Clark briefed approximately 50 representatives from a wide array of public and private sector engineering groups regarding the Chinese invitation. He challenged the representatives to respond as they saw fit and set a target date of July 15 for reply.

The group rose to the challenge. Initial organizing efforts were done by a core group composed of representatives of the U.S. Department of the Interior, American Consulting Engineers Council, National Council for U.S./China Trade, and private engineering firms. All interested parties were invited to donate resources to the effort, with no promise of return on their investment. The official title for the group that evolved was "The U.S. Three Gorges Working Group"—but William Clark also chose to christen the effort as

"Team America," reflecting the genuine patriotic spirit motivating the group on behalf of the nation's best interests.

Private and public groups contributing to the proposal furnished an estimated $1.5 million to $3 million in human, financial, and in-kind resources to this unique effort, which one participant described as the highlight of his career.

Participating firms and agencies called in their top people, many from overseas assignments, to work on a job with a sense of mission for the nation, a job where top managers rolled up their sleeves, hammered out figures, and worked past old rivalries and differences.

The team was composed of high-level executives such as chief executive officers, vice-presidents, and agency heads from public and private groups often known as competitors rather than cooperators. Side by side they shared their expertise in engineering design, construction, management, and financial and economic fields. Most of the participants had 30 years' experience in large-scale dam and hydroelectric power projects.

Representatives of these private sector firms made up the team:

- *Guy F. Atkinson Company*
- *Bechtel Civil and Minerals Engineering, Inc.*
- *Coopers and Lybrand*
- *Merrill Lynch Capital Markets*
- *The Morgan Bank*
- *Morrison Knudsen Corporation*
- *Stone and Webster Engineering Corporation*

The federal government's contribution came from services provided by the U.S. Department of the Interior's Bureau of Reclamation and the U.S. Army Corps of Engineers.

Each party made an offering. With 85 years' experience in design and construction of major water resources projects, the Bureau of Reclamation furnished approximately 20 experts in various fields to advance the proposal. Access to vital, existing data was possible through working agreements between the bureau and China and through bureau engineers who were then working at the Three Gorges site. The Corps of Engineers, one of the few existing bases of knowledge in the United States for lock design, provided invaluable expertise. The private firms contributed experience in preparing proposals and overall know-how on getting a job done on time and within budget.

An office was set up in Washington, D.C., as the base of operations. The leader of the private sector parties moved to the city for the six-week assignment, while most other participants commuted from their offices around the country. Work days often became work nights as the group propelled itself from its first meeting June 10 to the mid-July target date.

The tight deadline proved to be a great motivator, prompting the team to adopt more flexible, creative working methods than typically used in industry and government. Uncommon events often demand uncommon approaches, and one participant commented that, to his knowledge, a joint public-private effort of this magnitude had never before been attempted.

The executives were called on to use all the knowledge and abilities, both technical and managerial, that have made them successful in their organizations. The co-leaders, one from private industry and one from a federal agency, found they could not manage the group members as they would their own employees. Without the power conveyed by their respective organizations, they had to exercise personal skills to motivate the group to accept, support, and carry out shared objectives. The individuals practiced their interpersonal communication skills by offering constructive critiques as work progressed. Management books on the bestseller list talk about cases such as this that bring out the best in managers to build teams, integrate diverse talents, and manage disputes in pursuit of a first-class product.

Managerial skills were also required to address the unique organizational structure within the People's Republic of China relating to design, construction, and management of existing and planned water resource projects. The Chinese government had encountered substantial difficulties in building the Gezhouba Project on the Yangtze River downstream from the proposed site of the Three Gorges Project, mainly due to their complex system of interrelated ministries. The U.S. team worked on devising a more effective, simplified mechanism to avoid a recurrence of these problems on Three Gorges.

The outcome of this intensive effort was a comprehensive proposal including an implementation plan and economic study leading to a financial plan—all completed on schedule and with a great deal of pride. The proposal recommended using China's own technical and human resources to the extent possible to help the nation develop a strong base of knowledge. The effort proposed would foster an unprecedented level of cooperation and technology transfer between U.S. private and public sectors and the People's Republic.

The proposal, with a summary volume in both Chinese and English, was presented to China's Vice Premier Li Peng on July 17 at a setting appropriate for the occasion, the bureau's massive Hoover Dam. Later, in China, the proposal was presented to Madame Minister Qian, head of the Ministry of Water Resources and Electric Power.

While the Chinese were quick to embrace the proposal in principle, the question remained as to who would fund a feasibility study on the project. The government-to-government effort initiated through U.S.-Chinese working agreements and furthered by William Clark's visit had opened the door to future alliances, but neither the U.S. government nor U.S. private industry was able to surmount the stumbling block posed by the estimated cost of $6 million to $8 million for the feasibility study. The total cost of constructing the project is anticipated to be approximately $8 billion.

In October 1985, the Canadian government signed an agreement with the Ministry of Water Resources and Electric Power for joint participation in a feasibility study. The agreement includes a grant from Canada to China to fund the work of Canadian engineers. The cost of the study is estimated at $7.5 million to $8.3 million, and the anticipated completion date was December 1987.

A number of high-level Chinese officials have publicly stated that the Three Gorges Project will be built, but decline to establish specific time frames. Outside analysts predict that work will proceed when major issues are resolved, such as project financing, appropriate height of the dam, and environmental concerns.

When asked if they would do it again, executives involved with Team America answer with a resounding "yes." The participants view the experience as a positive one and a challenge from technical, managerial, and political standpoints. The hard hours may have temporarily exacted a toll, but the long-term payoff is an enduring sense of satisfaction on a personal and a professional level.

One spin-off of the team's work was exploding the stereotypes surrounding government workers in relation to their private sector counterparts. In the trenches, the team members found that talent, determination, and professionalism exist in many places. The involvement of experienced senior professionals from the public and private sectors was the key ingredient in producing a quality product on time.

The invitation from the People's Republic of China to prepare this proposal is an indicator of their respect for the technical and professional expertise found in U.S. private industry and government. Through other joint projects, the country's best human resources can be melded for a variety of purposes, including technology transfer to help other nations achieve their goals, and enhancing the position of the United States in international competition.

7
Needed: Institutional Structure to Promote Global Enterprise

The U.S. construction industry consists of 5.5 million individuals employed in 1.2 million firms, myriad professions and trades, and a variety of organizations representing these individuals. These many participants share common interests and concerns about the general health of the U.S. economy. While only a small fraction of these participants are active in the international construction market, they recognize the implications of U.S. weakness in this market, and they can understand the opportunities that technological leadership offers.

The committee has noted the high-level government focus for construction policy and export activity that some countries have established. The committee has noted as well the support for construction research and the close public-private partnership that industry in some other countries enjoys. Finally, the committee has noted the needs for the United States to catch up in its research and development, professional training, and pursuit of innovation in construction.

ORGANIZED FOCUS OF DIVERSE INTERESTS

The committee concludes that a more effective way is needed to bring together on a continuing basis the many diverse private and public interests in the U.S. construction industry, to resolve inevitable conflicts of opinion among these interests, and thereby to give the industry stronger voice in the national policy forum. Professional societies and trade associations, such as the American Institute of Architects, American Society of Civil Engineers, and Association of General Contractors, currently play an important role in representing the interest of their members, but there is no effective means to bridge the differences among groups. A solid institutional focus is needed to provide greater unity within the industry and to facilitate concerns and coordinated action. Existing institutions could be given expanded mandates to play such a role, but new institutions may be needed.

The committee has found it difficult to understand why the United States, as a nation, was unable or unwilling to allocate the funds to support its already substantial private investment in the Three Gorges Project, while its much smaller northern neighbor found the allocation to be in its national interest. At $8 million the amount is meagre when compared to government spending on any number of programs to support various other sectors of the U.S. economy.

A trade agreement signed with Japan in early 1988 offers possible resolution of the problems already described regarding U.S. construction industry activity in the Japanese market. However, in the heat of long-running negotiations, the United States appears to have lost sight of its main interest: the technically advanced segment of the construction market. Apparent access to a range of smaller projects that are largely labor and materials intensive will not only hold little attraction for U.S. firms, but will then hurt future U.S. prospects by giving the appearance that the nation is not serious about global enterprise. Both sides in the agreement are reported to hold a "show-me" attitude (*Engineering News Record,* April 7, 1988, pp. 12–13).

While the U.S.-Japanese trade negotiations proceeded, the French government-sponsored consulting firm Aeroports de Paris, which had been hired to evaluate proposed designs for passenger terminals at Kansai International Airport, invested its resources in preparation of its own alternative proposal. Its innovative plan swayed the airport authority's opinion and led to a new design competition, creating an opportunity for which French designers (and ultimately, constructors and equipment suppliers as well) now appear to hold a distinct advantage.

The committee felt these cases are not unusual, but rather are examples of a pattern of poorly focused attention and seeming lack of interest in U.S. construction within an increasingly global marketplace. Further analysis is needed to define the pattern more clearly and to identify what should be done to correct what is, in the committee's view, a problem that will have increasingly serious consequences for the nation's well-being. Nevertheless, it is readily apparent that the United States lacks the means to bring together public and private groups to offer the best of U.S. construction skills and technology in world markets. The institutional structure is needed to facilitate the cooperation illustrated in the pursuit of China's Three Gorges Project, and then to follow through with the support needed to strengthen the nation's ability to compete or to develop cooperative ventures with international partners.

The institutional structure could take any number of forms:

- There could be at the apex a federal government agency responsible for supporting international and domestic construction enterprise. This government office could propose policy initiatives for legislative action and coordinate government activity that influences the construction industry.
- There could be a quasi-governmental organization that would assemble U.S. construction experts from a variety of firms and government to work with counterpart organizations found in other countries. This organization could act to represent U.S. interests in international competition for major design and construction projects.
- There could be a unit associated with government, but not an agency of government, that would monitor the performance of the U.S. construction industry and government policies that influence that performance. This unit would serve as an objective observer and forum for identifying problems and defining options for solving these problems.

Perhaps some combination of such organizations is appropriate. However, this institutional focus is needed, its exact form must be determined, and the committee recommends that study should proceed.

ATTITUDE OF OPPORTUNITY

The design and construction industries in Western societies (and in Japan) believe they are faced with declining markets because of stable populations. Other countries have targeted the U.S. market because it is so open and large that it seems a natural way to gain business that will offset their own shrinking volume. However, an international cooperative effort to advance the technology of infrastructure could create whole new markets for urban and interurban systems with higher-performance characteristics.

Development of advanced infrastructure is a challenge worthy of cooperative international effort. It will be difficult to structure these developments to match the performance requirements of a society utilizing advanced science and technology, and make more than incremental improvements to the present modal technologies. In the developing part of the world, which is experiencing the most rapid urbanization, the challenge is to develop technological applications appropriate to specific-case requirements, rather than to impose solutions produced for industrial nations.

There are two reasons for the United States to do more toward advancing the technology of infrastructure. The nation would benefit within its own borders from new and higher-performance systems, and it could also enhance the opportunity for marketing its technology on a global basis. This committee recognizes the urgency of maintaining and extending the existing networks of public works that underlie the nation. However, the United States also needs to develop new and higher-performing technologies to enhance our competitive position in the world.

The committee recommends that action is needed at a national level to deal with the issues of liability and societal risk aversion that discourage large companies from introducing potentially innovative technologies. Increased government commitment to research and innovation are needed, through programs to apply new technology as well as through financial support of construction research and development.

RESEARCH AND DEVELOPMENT AND INNOVATION

The degree to which research and development activity will lead directly to innovation in infrastructure or in construction in general may be a subject of debate, but it is apparent to the committee that the United States is currently spending too little on construction research and development. Means must be found to enhance the apparent advantages that private companies can realize from this investment, for example, through changes in tax policy, risk sharing on government-sponsored projects, or modification of procurement procedures to support purchase of innovative design and materials applications.

Because infrastructure is built primarily for government clients and in large investment increments, policies to encourage research and development—and innovation—may most easily be developed in this area. The committee recommends that further work be undertaken to define and implement these policies.

BUILDING FOR TOMORROW

The nation is faced with a challenge to build for tomorrow. The strategic and commercial rewards of meeting this challenge will be surpassed only by the rewards of improved quality of life for the citizens of an increasingly global economy.

Competitive position is the topic with which this committee started, but it is not the proper end. Technological advance in construction of buildings and infrastructure can bring enhanced productivity and improved quality of life to all nations, yielding in due course increased business opportunity for foreign firms as well as U.S. industry. This is opportunity on a global scale, and the U.S. construction industry can play a leadership role in the enterprise. Building for the future is the best possible course for U.S. construction in an increasingly global economy.